Inherent Safety at Chemical Sites

Inherent Safety at Chemical Sites

Reducing Vulnerability to Accidents and Terrorism Through Green Chemistry

Paul T. Anastas
Teresa and H. John Heinz III Chair in Chemistry for the Environment
School of Forestry and Environmental Studies, Department of Chemistry,
Department of Chemical and Environmental Engineering, School of
Management, Yale University, New Haven, CT, USA

David G. Hammond
Senior Scientist, Aquagy, Inc., Berkeley, CA, USA

ELSEVIER
AMSTERDAM • BOSTON • HEIDELBERG • LONDON
NEW YORK • OXFORD • PARIS • SAN DIEGO
SAN FRANCISCO • SINGAPORE • SYDNEY • TOKYO

Elsevier
Radarweg 29, PO Box 211, 1000 AE Amsterdam, Netherlands
The Boulevard, Langford Lane, Kidlington, Oxford OX5 1GB, UK
225 Wyman Street, Waltham, MA 02451, USA

Notices
Knowledge and best practice in this field are constantly changing. As new research and
experience broaden our understanding, changes in research methods or professional practices,
may become necessary.

Practitioners and researchers must always rely on their own experience and knowledge in
evaluating and using any information or methods described herein. In using such information or
methods they should be mindful of their own safety and the safety of others, including parties for
whom they have a professional responsibility.

To the fullest extent of the law, neither the Publisher nor the authors, contributors, or editors,
assume any liability for any injury and/or damage to persons or property as a matter of products
liability, negligence or otherwise, or from any use or operation of any methods, products,
instructions, or ideas contained in the material herein.

ISBN: 978-0-12-804190-1

British Library Cataloguing-in-Publication Data
A catalogue record for this book is available from the British Library

Library of Congress Cataloging-in-Publication Data
A catalog record for this book is available from the Library of Congress

For Information on all Elsevier publications
visit our website at http://store.elsevier.com/

This book has been manufactured using Print On Demand technology.

CONTENTS

"This book is dedicated to those lives that have been lost or damaged in tragedies that could have been avoided through the use of Green Chemistry."

ACKNOWLEDGMENTS

The material presented and cases profiled in this book are primarily the product of hard work by a great many people, many of whom it is impossible to recognize adequately or comprehensively because they are originally sourced from government reports. Where known, appropriate credit has been given via the references section below. Special acknowledgment and credit is due Paul Orum, the author of "Preventing Toxic Terrorism. How Some Chemical Facilities are Removing Danger to American Communities." We also thank David Emmerman of Yale University and Jennifer Young of the Green Chemistry Institute for her review and helpful comments.

CHAPTER 1

Introduction

Since its inception as a conscious strategy in the early 1990s, green chemistry has gained recognition as a reliable and cost effective means for reducing the environmental impacts of industry. But a less anticipated side benefit of green chemistry methods has been that they also help to protect America's extensive chemical infrastructure from the threats of terrorism. A company's drive to save power, reduce waste, or use and store smaller quantities of hazardous chemicals will dictate modifications that also tend to reduce vulnerability to catastrophic accidents perpetrated by would-be saboteurs. At a time when concern over terrorism is running high, decision-makers in the chemical industry are wisely examining how they can incorporate green chemistry techniques to decrease their exposure to risk.

Across the United States, approximately 15,000 chemical plants, manufacturers, water utilities, and other facilities store and use extremely hazardous substances that would injure or kill employees and residents in nearby communities if suddenly released. Approximately 125 of these facilities each put at least 1 million people at risk; 700 facilities each put at least 100,000 people at risk; and 3000 facilities each put at least 10,000 people at risk, cumulatively placing the well-being of more than 200 million American people at risk,[1] in many cases unnecessarily. The threat of terrorism has brought new scrutiny to the potential for terrorists to deliberately trigger accidents that until recently the chemical industry characterized as unlikely worst-case scenarios. Such an act could have even more severe consequences than the thousands of accidental releases that occur each year as a result of ongoing use of hazardous chemicals.

The Department of Homeland Security and numerous security experts have warned that terrorists could turn hazardous chemical facilities into improvised weapons of mass destruction. As far back as 1999, the Agency for Toxic Substances and Disease Registry warned

Inherent Safety at Chemical Sites. DOI: http://dx.doi.org/10.1016/B978-0-12-804190-1.00001-X

that industrial chemicals provide terrorists with "...effective, and readily accessible materials to develop improvised explosives, incendiaries and poisons."[2,3]

The prospect of a deliberate act targeting chemical production or storage sites is frightening for its potential—via release and dispersal of noxious chemicals into our air, soil, and waterways—to harm people, property, and resources. Furthermore, the long-term impacts and consequences of such an incident could go far beyond the direct initial damage of a hostile strike.

Fortunately, ingenuity has bred novel ways to lower our vulnerability to terrorist attack, spawning strategies that supersede the mere strengthening of physical barriers. Whereas fences, walls, alarms, and other physical safety measures will always have some possibility of failure—particularly when the enemy wields weapons like airplanes and bombs—the wholesale replacement of hazardous chemicals with benign and inherently safer, or "greener" materials is a preventative measure that is guaranteed to provide fail-safe results. A hazardous chemical that is no longer present can no longer be turned into a weapon to be used against you. It is estimated that employing alternative chemicals at the nation's 101 most hazardous facilities could improve the security of 80 million Americans.[4]

Experts in the field of risk assessment are, therefore, concluding that green chemistry methods, though initially motivated by environmental or sometimes economic concerns, also offer the important additional benefit of decreasing our exposure to the threats of terrorism.

This book briefly introduces the concepts of green chemistry, and shows the various ways that a green approach to chemical design, production, and management is not only good for the planet, but also serves to protect people and infrastructure from terrorist acts. Specific examples and case studies are cited to illustrate what has been done to advance this cause, and offer guidance to those decision-makers who similarly aspire to greater safety and security for the people and resources they manage.

By focusing primarily on tangible case studies, we describe here the green chemistry innovations implemented by each company or facility. Where possible, we include details comparing the new technology to previous or conventional methods, and broadly quantify the improvements in terms of hazardous chemicals avoided or people protected. Although the specific details of each chemical process cannot be guaranteed to be accurate, they are presented in good faith and to the best of our knowledge; we encourage interested parties to seek further information directly from the relevant parties or from collaborative industry groups.

1.1 WHAT EXACTLY IS GREEN CHEMISTRY?

Green chemistry is the design of chemical products and processes in a manner that reduces or eliminates the use and generation of hazardous substances.[5] The term "hazardous" is employed in its broadest context to include physical (e.g., explosion, flammability), toxicological (e.g., carcinogenic, mutagenic), and global (e.g., ozone depletion, climate change) considerations. Green chemistry is an approach to the synthesis, processing, and use of chemicals that inherently reduces risks to humans and the environment.[6] A concern for both use and generation of hazardous substances is essential because it ensures that the chemist or designer address complete life cycle considerations.[7]

Typical modifications that have proven fruitful in furthering the cause of green chemistry include replacing particularly hazardous chemicals with less problematic alternatives, minimizing the amount of hazardous material needed for a reaction by combining it with a catalyst to increase the effective yield, and manufacturing material on-site or on-demand so as to minimize the amount stored, handled, and transported.[8]

Unlike add-on safety measures such as barriers, locks, employee training, and emergency response systems—which can never be 100% reliable because there is always some potential for a breach or accident—green chemistry techniques that result in a fundamental change in process or materials offer permanent, ensured improvements.

In the words of Trevor Kletz, a pioneer of inherently safe chemical engineering, "what you don't have, can't leak."[9] Likewise, what you don't have can't be made the target of a terrorist attack.

The key design principles that drive innovation in the field of green chemistry have been summarized[10] as:

Principles of Green Chemistry

1. **Prevention**

   It is better to prevent waste than to treat or clean up waste after it has been created.

2. **Atom Economy**

   Synthetic methods should be designed to maximize the incorporation of all materials used in the process into the final product.

3. **Less Hazardous Chemical Syntheses**

   Wherever practicable, synthetic methods should be designed to use and generate substances that possess little or no toxicity to human health and the environment.

4. **Designing Safer Chemicals**

   Chemical products should be designed to affect their desired function while minimizing their toxicity.

5. **Safer Solvents and Auxiliaries**

   The use of auxiliary substances (e.g., solvents, separation agents) should be made unnecessary wherever possible and innocuous when used.

6. **Design for Energy Efficiency**

   Energy requirements of chemical processes should be recognized for their environmental and economic impacts and should be minimized. If possible, synthetic methods should be conducted at ambient temperature and pressure.

7. **Use of Renewable Feedstocks**

   A raw material or feedstock should be renewable rather than depleting whenever technically and economically practicable.

8. **Reduce Derivatives**

   Unnecessary derivatization (use of blocking groups, protection/deprotection, temporary modification of physical/chemical processes) should be minimized or avoided if possible, because such steps require additional reagents and can generate waste.

9. **Catalysis**

   Catalytic reagents (as selective as possible) are superior to stoichiometric reagents.

10. **Design for Degradation**

    Chemical products should be designed so that at the end of their function they break down into innocuous degradation products and do not persist in the environment.

11. **Real-time analysis for Pollution Prevention**
    Analytical methodologies need to be further developed to allow for real-time, in-process monitoring and control prior to the formation of hazardous substances.
12. **Inherently Safer Chemistry for Accident Prevention**

    Substances and the form of a substance used in a chemical process should be chosen to minimize the potential for chemical accidents, including releases, explosions, and fires.

1.2 RECENT TRENDS IN GREEN CHEMISTRY

Research contributing to the greening of chemistry is conducted around the globe by experts and innovators in diverse areas, including polymers, solvents, catalysts, renewables, bio-based materials, water treatment, and analytical methods, and has resulted in a wide variety of interesting new products and processes. What follows is a small sampling to illustrate the breadth and far-reaching impact of the innovative contributions to recently come from the field of green chemistry.

1.2.1 Synthetics from Glucose

Chemical intermediates, such as catechol and adipic acid, used in the manufacture of nylon-6,6, polyurethane, lubricants, and plasticizers are normally derived from petroleum-based benzene and toluene. Airborne benzene causes cancer and leukemia[11,12]; toluene leads to brain, liver, and kidney damage, and debilitates capacity for speech, vision, and balance.[13,14] Researchers at Michigan State have developed a green method for biosynthesizing catechol and adipic acid from glucose, rather than from benzene and toluene, using genetically altered *E. coli*.[15–17]

1.2.2 Chromium- and Arsenic-Free Wood Preservative

As of 2002, more than 95% of pressure-treated wood in the United States was treated with chromated copper arsenate (CCA). In 2003, the U.S. EPA prohibited the use of CCA-treated wood in residential settings. CCA poses a public health threat through its production, transportation, use, and disposal, and is especially harmful to children, who are more susceptible, and readily contact CCA-treated wood in playgrounds, decks, and picnic tables. Chemical Specialties, Inc. has developed an alkaline copper quaternary (ACQ) wood preservative that does not create any hazardous waste in its production and treatment. If fully adopted, ACQ will eliminate 90% of the 44 million pounds of arsenic currently used in the United States, as well as 64 million pounds of hexavalent chromium. None of the ACQ constituents are considered carcinogens by the World Health Organization.

1.2.3 Greenlist™ Process to Reformulate Consumer Products

SC Johnson (SCJ) formulates and manufactures consumer products including a wide variety of products for home cleaning, air care, personal care, insect control, and home storage. SC Johnson developed Greenlist™, a system that rates the environmental and health effects of the ingredients in its products. SC Johnson is now using Greenlist™ to reformulate many of its products to make them safer and more environmentally responsible. For example, "Greenlisting" Saran Wrap® resulted in converting it to low-density polyethylene, eliminating the use of nearly 4 million pounds of polyvinylidene chloride (PVDC) annually. In another example, SCJ used the list system to remove a particular volatile organic compound (VOC) from Windex®. They developed a novel formula containing amphoteric and anionic surfactants, a solvent system with fewer than 4% VOCs, and a polymer for superior wetting. Their formula cleans 30% better and eliminates over 1.8 million pounds of VOCs per year. Through Greenlist™, SCJ chemists and product formulators around the globe now have instant access to environmental ratings of potential product ingredients.

1.2.4 Greener Chemicals for Medical Imaging

A photothermographic technology developed by Imation, Inc. for their DryView™ Imaging Systems replaces silver halide photographic films for medical imaging. This technology also replaces all of the photographic developer and fixer solutions containing toxic chemicals, such as hydroquinone, silver, and acetic acid. Silver halide photographic films are processed by being bathed in a chemical developer, soaked in a fix solution, washed with clean water, and finally dried. The developer and fix solutions contain toxic chemicals, such as hydroquinone, silver, and acetic acid. In the wash cycle, these chemicals, along with silver compounds, are flushed from the film, and become part of the waste stream. The resulting effluent amounts to billions of gallons of liquid waste each year. During 1996, Imation delivered more than 1,500 DryView™ medical laser imagers worldwide, representing 6% of the world's installed base. These units are responsible for eliminating the annual disposal of 192,000 gallons of developer, 330,000 gallons of fixer, and 54.5 million gallons of contaminated water. As future systems are placed, the reductions will be even more dramatic.

1.2.5 100% CO_2 as Blowing Agent for Polystyrene Foam Packaging

This process from Dow Chemical for manufacturing polystyrene foam sheets uses 100% carbon dioxide (CO_2) as a blowing agent, eliminating 3.5 million pounds per year of traditional blowing agents, which deplete the ozone layer, are greenhouse gases, or both. The Dow Chemical Company will obtain CO_2 from existing commercial and natural sources that generate it as a byproduct, ensuring no net increase in global CO_2. Unlike traditional blowing agents, the new 100% CO_2 blowing agent will not deplete the ozone layer, will not contribute to ground level smog, and will not contribute to global warming.

1.2.6 Environmentally Safe Marine Anti-Foulant

Rohm and Haas have introduced Sea-Nine™, a new anti-foulant, to replace environmentally persistent and toxic organotin anti-foulants, such as tributyltin oxide (TBTO). Currently, fouling costs the shipping industry approximately $3 billion a year in increased fuel consumption needed to overcome hydrodynamic drag; increased fuel consumption subsequently contributes to pollution, global warming, and acid rain. Sea-Nine™ anti-foulant degrades extremely rapidly with a half-life of one day in seawater and one hour in sediment.

1.2.7 Green Synthesis for Active Ingredient in Diabetes Treatment

Merck has discovered a more efficient catalytic synthesis for sitagliptin, a chiral β-amino acid derivative that is the active ingredient in their new treatment for type 2 diabetes, Januvia™. This revolutionary synthesis creates 220 pounds less waste for each pound of sitagliptin manufactured, and increases the overall yield by nearly 50%. Over the lifetime of Januvia™, Merck expects to eliminate the formation of at least 330 million pounds of waste, including nearly 110 million pounds of aqueous waste.

Merck used a first-generation synthesis of sitagliptin to prepare over 200 pounds for clinical trials. With modifications, this synthesis could have been a viable manufacturing process, but it required eight steps including a number of aqueous work-ups. It also required several high-molecular-weight reagents that were not incorporated into the final molecule and, therefore, ended up as waste.

While developing a second-generation synthesis for sitagliptin, Merck researchers discovered a completely unprecedented transformation: the asymmetric catalytic hydrogenation of unprotected enamines. In collaboration with Solvias, a company with expertise in this area, Merck scientists discovered that hydrogenation of unprotected enamines using rhodium salts of a ferrocenyl-based ligand as the catalyst gives β-amino acid derivatives of high optical purity and yield. This new method provides a general synthesis of β-amino acids, a class of molecules well known for interesting biological properties.

1.2.8 Ionic Liquids Dissolve Cellulose for Reconstitution into Advanced New Materials

University of Alabama's Professor Rogers has invented a method that allows cellulose to be (1) chemically modified to make new biorenewable or biocompatible materials; (2) mixed with other substances, such as dyes; or (3) simply processed directly from solution into a formed shape.[18]

Major chemical companies are currently making tremendous strides towards using renewable resources in biorefineries. In a typical biorefinery, the complexity of natural polymers, such as cellulose, is first broken down into simple building blocks (e.g., ethanol, lactic acid), then built up into complex polymers. If one could use the biocomplexity of natural polymers to form new materials directly, however, one could eliminate many destructive and constructive synthetic steps. Professor Rogers and his group have successfully demonstrated a platform strategy to efficiently exploit the biocomplexity afforded by one of nature's renewable polymers, cellulose, potentially reducing society's dependence on nonrenewable petroleum-based feedstocks for synthetic polymers. No one had exploited the full potential of cellulose previously, due in part to the shift towards petroleum-based polymers since the 1940s, nor the difficulty in modifying the cellulose polymer properties, and the limited number of common solvents for cellulose.

Professor Rogers's technology combines two major principles of green chemistry: developing environmentally preferable solvents, and using biorenewable feedstocks to form advanced materials. Professor Rogers has found that cellulose from virtually any source (fibrous, amorphous, pulp, cotton, bacterial, filter paper, etc.) can be dissolved readily and rapidly, without derivatization, in a low-melting ionic liquid (IL), 1-butyl-3-methylimidazolium chloride, by gentle heating (especially with microwaves).

IL-dissolved cellulose can easily be reconstituted in water in controlled architectures (fibers, membranes, beads, flocs, etc.) using conventional extrusion spinning or forming techniques. By incorporating functional additives into the solution before reconstitution, Professor Rogers can prepare blended or composite materials. The incorporated functional additives can be either dissolved (e.g., dyes, complexants, other polymers) or dispersed (e.g., nanoparticles, clays, enzymes) in the IL before or after dissolution of the cellulose. With this simple, noncovalent approach, one can readily prepare encapsulated cellulose composites of tunable architecture, functionality, and rheology.

The IL can be recycled by a novel salting-out step or by common cation exchange, both of which save energy compared to recycling by distillation. Professor Rogers's current research is aimed at improved, more efficient, and economical syntheses of this particular IL, and studies of its toxicology, engineering process development, and commercialization.

As of 2013, the researchers were engaged in market research and business planning leading to the commercialization of targeted materials, either through joint development agreements with existing chemical companies or through the creation of small businesses. Green chemistry principles will guide the development work and product selection. For example, targeting polypropylene- and polyethylene-derived thermoplastic materials for the automotive industry could result in materials with lower cost, greater flexibility, lower weight, lower abrasion, lower toxicity, and improved biodegradability, as well as significant reductions in the use of petroleum-derived plastics. ILs remain expensive and energy intensive, but the researchers believe their costs will go down with time.[19]

Professor Rogers's work combines a fundamental knowledge of ILs as solvents with a novel technology for dissolving and reconstituting cellulose and similar polymers. Using green chemistry principles to guide process development and commercialization, he envisions that his platform strategy can lead to a variety of commercially viable advanced materials that will obviate or reduce the use of synthetic polymers.

CHAPTER 2

Accident Vulnerability and Terrorist Threats to the Chemical and Related Industries

The chemicals reduced or eliminated in many successful examples of green chemistry (such as those profiled in the previous chapter) are not necessarily substances of particular interest to terrorists because they may not pose an immediate threat to people, or that threat may be very limited in scale. Nevertheless, there is no question that many chemical facilities do in fact constitute tempting targets for saboteurs wishing to cause harm to large numbers of the population.

In May of 2002 a truck loaded with explosives and rigged for detonation from a cell phone was driven into Israel's largest fuel depot located near densely populated Tel Aviv. Flames from the exploding truck were extinguished before they could spread to nearby tanks containing millions of gallons of fuel, but the narrowly averted catastrophe at a storage and distribution point situated in the middle of a residential neighborhood and furthermore, unnervingly close to security and military intelligence installations illustrated the vulnerability that chemical sites pose to millions of civilians in urban areas worldwide. Even amid prior threats to the fuel depot and increased security, guards who checked the truck at the entrance failed to notice the bomb attached to its chassis.[20]

Chemicals that ultimately pose the greatest threat to public safety are those that are especially explosive or volatile. An FBI report that analyzed statistics of domestic terrorist attacks found that 93% of the incidents involved the use of explosives or incendiaries.[13] Perhaps the worst-case scenario involves a sudden and uncontrolled release of toxic gas that is heavier than air, and moves along the ground, spreading downwind as an invisible yet deadly plume.

Some authorities are convinced that Mohamed Atta, believed to have been a ringleader of the September 11 terrorists, had evaluated at least one Tennessee chemical storage facility—housing dozens of round

Inherent Safety at Chemical Sites. DOI: http://dx.doi.org/10.1016/B978-0-12-804190-1.00002-1

steel tanks, flanked by towering smokestacks, and surrounded by hundreds of rail tanker cars—as a potential target, inquiring insistently about the contents of the tanks and rail cars. Coincidentally, the plant's owner, Intertrade Holdings, had recently stopped storing sulfuric acid and other hazardous chemicals in the tanks in preparation for closing the plant's acid manufacturing operation. Another individual suspected to have been an associate of the 9/11 terrorists had acquired a license to haul hazardous materials in Michigan.

At the time of this writing, there are very few instances of a U.S. chemical facility being successfully attacked by terrorists,[21,22,23] but heightened concern over the scope and frequency of deliberate strikes has forced consideration of how to best prevent the potentially staggering consequences of such an event.

One place to begin in assessing which chemicals pose the most dangers in the event of a terrorist attack is to look at the chemicals that have most often been involved in past industrial accidents. Information gathered through the EPA's Risk Management Planning (RMP) program (explained further in Tables 2.1 and 2.2) has been compiled to reflect the number of accidents nationwide between 1994 and 2000, and the industries in which these accidents occurred. The findings give us a valuable window into the risks inherent in production and handling of key industrial chemicals.

Unfortunately, accidents with chemicals are common, and the data in these tables is by no means comprehensive, intended, rather, to give a picture into the relative risks associated with different chemicals and industries. The National Response Center—the federal agency to which oil and chemical companies report oil and chemical spills— estimates that each year there are more than 25,000 fires, spills, or explosions involving hazardous chemicals, with about 1000 of these events involving deaths, injuries, or evacuations[1,25] (Figure 2.1).

2.1 CHEMICALS VULNERABLE TO TERRORISM OR ACCIDENTS

Because there are numerous chemical compounds prone to accidents— and, therefore, also to terrorist attack—we must prioritize those that pose the greatest risks. Criteria having the most influence include their prevalence by industry and geography, gross volumes used, severity of

Table 2.1 Chemicals That Most Frequently Create Accident Risks

Chemical	Number of Processes	Percentage of Total
Ammonia (anhydrous)	8343	32.5
Chlorine	4682	18.3
Flammable Mixtures	2830	11
Propane (industrial use)	1707	6.7
Sulfur Dioxide	768	3
Ammonia (aqueous 20% or more conc.)	519	2
Butane	482	1.9
Formaldehyde	358	1.4
Isobutane	344	1.3
Hydrogen Fluoride	315	1.2
Pentane	272	1.1
Propylene	251	1
Methane	220	0.9
Hydrogen	205	0.8
Isopentane	201	0.8
All Others	4139	16.1
Total	**25636**	**100%**

Note that four chemicals are present in nearly 70% of all processes reported to EPA's RMP.[24]

Table 2.2 Industries with the Most High-Risk Processes in EPA's RMP

Industry NAICS Code and Description	Number of Processes	Percentage of All RMP
42291 Farm Supplies Wholesalers	4409	28.84
22131 Water Supply & Irrigation	2059	13.47
22132 Sewage Treatment	1646	10.77
32411 Petroleum Refineries	1609	10.52
325199 All Other Basic Organic Chemical Manufacturing	655	4.28
42269 Other Chemical and Allies Products Wholesalers	607	3.97
49312 Refrigerated Warehousing and Storage Facilities	549	3.59
211112 Natural Gas Liquid Extraction	533	3.49
325211 Plastics Material and Resin Manufacturing	418	2.73
325188 All Other Basic Inorganic Chemical Manufacturing	358	2.34
49313 Farm Product Warehousing	345	2.26

(Continued)

Table 2.2 (Continued)		
Industry NAICS Code and Description	Number of Processes	Percentage of All RMP
32511 Petrochemical Manufacturing	321	2.1
454312 Liquefied Petroleum Gas Dealers	311	2.03
11511 Support Activities for Crop Production	302	1.98
311615 Poultry Processing	253	1.65
115112 Soil Preparation, Planting, and Cultivating	207	1.35
32512 Industrial Gas Manufacturing	205	1.34
325998 All Other Miscellaneous Chemical Product Manufacturing	193	1.26
325311 Nitrogenous Fertilizer Manufacturing	159	1.04
49311 General Warehousing and Storage Facilities	151	0.99
TOTAL	15,290	100%
Note that four industries account for more than 60% of processes reported to EPA's RMP.[24]		

Figure 2.1 A chemical plant in the industrial section of north Fort Worth explodes into flames in July 2005, sending toxic smoke hundreds of feet into the air. The blast and subsequent fire were fueled by a mixture of sulfuric acid, hydrochloric acid, ethanol, methanol, and isopropyl alcohol, with 30 different chemicals used and stored in tanks at the plant. Injuries from exposure to the fumes were apparently limited by fortuitously strong winds that helped to dissipate the plume relatively quickly.[26] Photo reprinted with permission, courtesy of David Bailey.

their effects when released, irreversibility of their effects if released, and the ready availability of less hazardous alternatives. Gleaning such information from the RMP tables above and a variety of other sources for accident, volatility, explosivity, and toxicity data, this report focuses on chemical compounds most likely to be targeted by terrorists, some of which are listed below.

Chemicals considered likely targets for terrorist attack based primarily on their high toxicity or volatility,[8] or their role in past chemical accidents:

1. Acrolein
2. Ammonia
3. Ammonium nitrate
4. Bromine
5. Chlorine
6. Cyanide and Hydrocyanic acid
7. Dioxin
8. Ethylene oxide
9. Formaldehyde
10. Hydrogen chloride and hydrochloric acid
11. Hydrogen fluoride and hydrofluoric acid
12. Hydrogen sulfide
13. Methyl isocyanate
14. Methyl mercaptan
15. Mononitrotoluene
16. Nitric acid
17. Nitric oxide
18. Nitrogen dioxide
19. Oil (contaminated water ways)
20. Phosgene
21. Propylene oxide
22. Sulfur
23. Sulfur dioxide, trioxide and sulfuric acid

The Chemical Emergency Preparedness and Prevention Office has estimated the zone of vulnerability under worst-case scenario conditions for facilities containing different hazardous substances. They conclude that for a facility containing toxic substances, the median distance from the facility to the outer edge of its vulnerable zone is 1.6 miles. Flammable substances have a worst-case scenario vulnerability zone whose median distance reaches 0.4 miles from the facility. However, many facilities reported vulnerability zones extending 14 miles from the facility (primarily for urban area releases of chlorine stored in 90-ton rail tank cars) and 25 miles (for rural releases of chlorine stored in 90-ton rail tank cars). Other chemicals for which the reported vulnerability zone equaled or exceeded 25 miles include anhydrous ammonia, hydrogen fluoride, sulfur dioxide, chlorine dioxide, oleum (fuming sulfuric acid),

Figure 2.2 More than 2000 residents were evacuated and 43 injured during the massive fires of December 2005 at the Hertfordshire oil depot outside of London, where 20 petrol tanks—each holding 3 million gallons of fuel— exploded. The blast blew doors off of houses in the surrounding area, sent flames hundreds of feet into the sky, and then burned for two and a half days. Authorities believe the incident to be an accident, but stated that the ferocity of the blaze destroyed all evidence and made it extremely hard for forensic experts to find out the cause.[27,28]

sulfur trioxide, hydrogen chloride, hydrocyanic acid, phosgene, propionitrile, bromine, and acrylonitrile.[24,26]

Chemical products technically represent a modest 2% of U.S. gross domestic products,[22] yet they are the foundation for a vast array of other manufactured goods, including plastics, fibers, drugs, paper, fabrics, cosmetics, and electronics, so disruptions to the chemical infrastructure can send lasting reverberations throughout the economy, and have severe impacts on our daily lives.

Aside from the risks posed by chemicals that are directly explosive or volatile, terrorist attacks might also target the supply chain of particular chemicals that are central and essential to our economy, comfort, or lifestyle (Figure 2.2).

CHAPTER 3

The Role of Green Chemistry in Reducing Risk

Numerous studies and institutions interpret and quantify the vulnerability of chemical sites, processes, and transportation methods to the varied threats of mechanical failure, human error, industrial accident, natural disaster, vandalism, theft, or terrorism, including the U.S. Chemical Safety and Hazard Investigation Board. As a result, most facilities have instituted a combination of voluntary and mandatory security measures to consistently improve their safety record. Nevertheless, it is an incontrovertible fact that no amount of security guards, fences, alarms, or containment structures can entirely eliminate risk at a site that produces, uses, or stores hazardous material. In contrast, when the chemists and engineers responsible for industrial process design seek to modify the process itself, inherently safer conditions can be permanently and irreversibly built into the chemical industry and its facilities.

In practical terms, some of the green chemistry approaches offering most promise for decreasing vulnerability to terrorist attacks include:

1. Replacement of a hazardous ingredient in the chemical synthesis process
2. On-site production of risk-heavy compounds (to minimize hazards associated with transportation)
3. On-demand production of risk-heavy compounds (to minimize amounts in storage)
4. Reducing reliance on those hazardous ingredients that cannot be replaced (e.g., by using catalysts to increase their effective yield)

3.1 AREAS WHERE GREEN CHEMISTRY HAS REDUCED RISK

Most industry efforts to date have focused on physical site-security measures that are unlikely to stop terrorists armed with airplanes and truck bombs.

Inherent Safety at Chemical Sites. DOI: http://dx.doi.org/10.1016/B978-0-12-804190-1.00003-3

Nevertheless, there are also hundreds of examples of facilities from a diverse range of industries that have successfully switched to safer chemical alternatives, including water utilities, manufacturers, power plants, waste management facilities, pool service companies, agricultural chemical suppliers, and the pharmaceutical and petroleum industries. These examples of green chemistry improvements that have already been implemented are proven as viable means to lower risk.

One 2006 survey conducted jointly by public interest, state, and environmental groups identified dozens of instances, where chemical dangers were dramatically reduced or successfully removed from their communities, and published these compelling findings[29]:

1. At least 284 facilities in 47 states have dramatically reduced the danger of a chemical release into nearby communities by switching to less acutely hazardous processes or chemicals, or by moving to safer locations.
2. As a result of these changes, at least 38 million people no longer live under the threat of a major toxic gas cloud from these facilities.
3. Eleven of these facilities formerly threatened more than one million people; a further 33 facilities threatened more than 100,000; and an additional 100 threatened more than 10,000.
4. Of respondents that provided cost estimates, roughly half reported spending less than $100,000 to switch to safer alternatives, and few spent over $1 million.
5. Survey respondents represented a range of facilities, small and large, including water utilities, manufacturers, power plants, service companies, waste management facilities, and agricultural chemical suppliers.
6. Facilities reported replacing gaseous chlorine, ammonia, and sulfur dioxide, among other chemicals.
7. The most common reasons cited for making changes included the security and safety of employees and nearby communities, as well as regulatory incentives and business opportunities.
8. Facilities cut a variety of costs and regulatory burdens by switching to less hazardous chemicals or processes. These facilities need fewer physical security and safety measures, and can better focus on producing valuable products and services.

It is heartening to realize, too, that most of these communities and facilities instituted changes that were relatively simple and unglamorous. Notably, many of the changes rely on common and available technologies, rather than new innovations. Thousands of additional facilities across a range of industries could make similar changes. They provide a template and set a precedent for improvements well within the economic and logistical reach of thousands more facilities that can profit from their experience.

One good example of a preventive response occurred at the Blue Plains sewage treatment plant, located in Blue Plains, Maryland, and serving Washington, DC.[1] The facility is situated across the Potomac River from the Pentagon, and before September 11, 2001, it housed multiple rail cars of chlorine and sulfur dioxide. Chlorine and sulfur dioxide are so volatile that the rupture of one full 90-ton tanker could spread a lethal cloud capable of killing people within 10 miles. From Blue Plains, such a swath could cover downtown Washington, DC, Anacostia, Reagan National Airport, and Alexandria.[30] Over the course of 8 weeks after September 11, authorities quietly removed up to 900 tons of liquid chlorine and sulfur dioxide, moving tanker cars at night under guard. "We had our own little Manhattan Project over here," Jerry N. Johnson, general manager of the D.C. Water and Sewer Authority, which runs the plant, told the Washington Post. "We decided it was unacceptable to keep this material here any longer."[30] The plant has since switched from volatile chlorine gas to sodium hypochlorite bleach, which is less volatile, and has far less potential for airborne off-site impact.

3.2 TRACKING TANGIBLE CHANGES THROUGH THE RISK MANAGEMENT PLANNING PROGRAM

One way to track the impact of green chemistry on reducing risk is to monitor companies' participation in the federal Risk Management Planning (RMP) program, which is administered by the U.S. Environmental Protection Agency (EPA). Approximately 15,000 facilities across the U.S. use hazardous industrial chemicals in quantities that trigger a requirement for regulation and periodic reporting to EPA. Each subject facility must prepare a Risk Management Plan that includes a hazard assessment, a prevention plan, and an emergency response plan.

Nearly 5000 of these facilities registered with the RMP have a maximum quantity of at least 100,000 pounds of a chemical considered extremely hazardous onsite—more than the amount released in the Bhopal, India disaster that killed thousands and left hundreds of thousands injured. At least 100 facilities each store the astounding figure of more than 30 million pounds of an extremely hazardous substance.[30] The potential for a catastrophic chemical release is widely distributed: every U.S. state, except Vermont, has at least one facility storing more than 100,000 pounds of an extremely hazardous substance.[31] The minimum threshold quantity necessitating registration under the RMP program ranges from 500–20,000 pounds, depending on the compound and its properties.[32]

The facilities must estimate how far a chemical could travel off-site in a worst-case release, along with the number of people living within the "vulnerability zone"—the area potentially affected by the release. These plans save lives, prevent pollution, and protect property by guiding companies in managing chemical hazards. Although not all people within the vulnerability zone would necessarily be injured by a single chemical release, the median number of people inside a facility's worst-case vulnerability zone is 1500 people. In addition to the risks faced by the general population, workers at every facility, and the emergency workers who would respond to an incident, are the most likely to be injured or killed in a chemical release.

As companies find and institute green alternatives to the hazardous chemicals regulated under the RMP program, they are relieved of the burden of reporting. Many of the examples presented in this report refer to facilities that have successfully switched from hazardous industrial chemicals to more benign alternatives, and as a consequence have freed themselves from the requirement to report to the RMP program. A very strong sampling of such changes was documented in an invaluable survey conducted by the Center for American Progress[29] (CAP) that gathered representative data from 284 diversified facilities in 47 states that, since 1999, have deregistered from the RMP program. Since the program's inception in 1999, there has been a notable decline in hazardous chemical facilities that report a vulnerability zone of more than 10,000 people, with the number of these high-hazard facilities declining by at least 544, from 3055 facilities to 2511.

As a result of these changes, more than 38 million Americans no longer live under the threat of a harmful toxic gas release from these facilities.[29] Eleven of these facilities formerly threatened more than one million people; another 33 facilities threatened more than 100,000; and an additional 100 threatened more than 10,000.

Terrorist threat heightens the risk presented by facilities that still have large vulnerability zones. However, the RMP program does not currently address the potential for a deliberate terrorist release of chemicals. Federal law does not require companies to assess readily available alternative chemicals and processes that pose fewer dangers.

3.3 WHY DO COMPANIES CHOOSE GREENER CHEMICALS OR PROCESSES?

Facility owners and managers most commonly give reasons of safety, security, regulatory requirements, and community expectations when asked to explain why they have chosen to switch to less hazardous chemicals or processes. After being presented with a variety of reasons for change, and instructed to check all explanations that apply, affirmative responses were given by the following percentage of respondents:

1. Concern over an accidental chemical release and improved safety 76%
2. Concern over terrorism and improved security 41%
3. Legal or regulatory requirements 37%
4. Meeting community expectations 20%
5. Improved operations efficiency or business opportunities 13%
6. Projected cost savings 12%
7. Other 10%
8. No answer 16%

3.4 COSTS AVOIDED WITH SAFER ALTERNATIVES

Plant and facility managers surveyed have identified a variety of costs and regulatory burdens that their facilities fully or partly eliminated as a result of switching to less hazardous substances or processes. Avoided costs mentioned in survey responses include the following:

1. Theft and theft prevention
2. Personal protective equipment (such as gas masks)
3. Safety devices (such as leak detection or scrubbers)
4. Safety inspections
5. Higher risk-group insurance premiums
6. Potential liability
7. Regulatory certifications, permits, and fees

8. Compliance staff
9. Certain chemical purchases
10. Specialized emergency response teams
11. Hazardous material safety training
12. Lost work time from chemical exposures
13. Chemical damage to infrastructure
14. Certain fire code requirements
15. Certain physical security measures
16. Unreliable chemical supply lines
17. Placards and material safety data sheets
18. Community notification
19. Evacuation and contingency plans
20. Background checks
21. Compliance with OSHA Process Safety Management
22. Compliance with EPA Risk Management Planning

The plant manager of the City of Vicksburg's Water Treatment Facility in Mississippi commented that "making changes was cheaper than complying with RMPs."[8]

It is interesting to note also that of the 59 respondents who changed to safer alternatives before September 11, 2001, 25% indicated security as a reason for making the change. Of the 225 respondents who changed to safer alternatives after Sept. 11, 2001, 45% indicated that increased security was a reason.

CHAPTER 4

Case Studies—Green Chemistry in Practice

Not all hazardous chemicals are equally hazardous. They differ notably in the severity of their effects and in the reversibility of those effects. Clearly, acutely lethal chemicals pose more concern for terrorist attack than do carcinogens, whose effects might take years to manifest themselves.

A rapidly moving plume of toxic gas is of much more immediate concern than a plume of contaminated groundwater. Nevertheless, an attack that causes little quantifiable death or damage may still prove very damaging to society's psyche, morale, and comfort level by causing widespread alarm and lasting disruption of normal activities. Sudden release of certain chronically toxic compounds can evoke potent fear and a passionate, panicked reaction among the public, and, therefore, still warrant attention in the context of reducing vulnerability to terrorist attack.

The case studies presented here showcase a selection of hazardous chemicals that potentially threaten public safety and security by posing a viable target for terrorist attack and unintended accidents, but that have been reduced or replaced by less hazardous alternatives in the situations profiled here. When available and relevant, details of each innovation, its methods, and benefits have been provided.

These examples of green chemistry at work will be instructive to those who seek ideas and guidance in identifying appropriate methods for greening their own workplace or community while simultaneously fortifying national security.

4.1 CHLORINE (CL$_2$)

Background: Chlorine (Cl$_2$) is a greenish-yellow, toxic gas with a strong odor that irritates the respiratory system. It is used in chemical manufacturing, bleaching, disinfection, and for purifying water and

Inherent Safety at Chemical Sites. DOI: http://dx.doi.org/10.1016/B978-0-12-804190-1.00004-5

sewage treatment. Acute exposure can severely burn the eyes and skin, causing permanent damage, and may cause throat irritation, tearing, coughing, nose bleeds, chest pain, fluid build-up in the lungs (pulmonary edema), and death. Chronic exposure can damage the teeth and irritate the lungs, causing bronchitis, coughing, and shortness of breath. A single high exposure can permanently damage the lungs.

Chlorine is a strong oxidizing agent that can react explosively, or form explosive compounds with many common materials, and react with flammable materials. It is only slightly soluble in water, but combines with it to form hypochlorous acid (HClO) and hydrochloric acid (HCl). The unstable HClO readily decomposes, forming oxygen-free radicals. Because of these reactions, water greatly enhances chlorine's oxidizing and corrosive effects. When oxidized, chlorine splits hydrogen from water, causing the release of nascent oxygen and hydrogen chloride, which produce major tissue damage. Alternatively, chlorine may be converted to hypochlorous acid, which can penetrate cells and react with cytoplasmic proteins to form N-chloro derivatives that can destroy cell structure. Symptoms may be apparent immediately or delayed for a few hours.[33]

Chlorine's most important use is as bleach in the manufacture of paper and cloth. It is also used widely as a chemical reagent in the synthesis and manufacture of metallic chlorides, chlorinated solvents, pesticides, polymers, synthetic rubbers, and refrigerants. Sodium hypochlorite—which is a component of commercial bleaches, cleaning solutions, and disinfectants for drinking water, wastewater purification systems, and swimming pools—releases chlorine gas when it comes in contact with acids.[33]

Because it is heavier than air, chlorine tends to hug the ground and accumulate at the bottom of poorly ventilated spaces, creating higher risk in certain workplace settings.

Chlorine was the second most frequently involved chemical in accidents documented for the RMP*Info database[24] (see Table 2.1). For example, in 2005, a switch error caused a 42-car train to collide with another parked train in South Carolina, puncturing a tank car full of chlorine. As a result of exposure to chlorine, nine people died; 250 were sent to the hospital and 5400 people had to be evacuated from the surrounding area.[34] Its abundance and affordability makes it an attractive agent for terrorists.

4.1.1 Chlorine in Water and Wastewater Treatment

A 2005 report issued by the Government Accountability Office (GAO) concluded that local wastewater treatment plants constitute a security threat because "chemicals used in wastewater treatment"—especially chlorine gas—are a "key vulnerability."[35]

Nationwide, there are more than 16,000 publicly owned wastewater systems that serve more than 200 million people, or about 70% of the total population.[36] Approximately 1150 wastewater facilities and 1700 drinking water plants are currently registered with the EPA's RMP program for extremely hazardous chemicals, primarily chlorine gas.[29]

According to EPA's report Pesticide Industry Sales and Usage, 1.56 billion pounds of chlorine were used as a disinfectant of potable water and wastewater in 2001. Another one billion pounds of chlorine were used as a disinfectant for recreational water. The report recommends the replacement of chlorine with less hazardous chemicals or practices.[37]

Approximately 114 wastewater facilities and 93 drinking water plants have reported switching to less acutely hazardous chemicals.[29] These facilities generally replaced chlorine gas with liquid chlorine bleach (sodium hypochlorite) or ultraviolet light. Some now generate bleach on-site in a dilute solution.

The GAO report found that adoption of alternatives has resulted in net savings. For example, the Blue Plains Wastewater Treatment Plant in Washington, D.C. employed around-the-clock police units prior to replacing its chlorine gas treatment process, realizing a savings when they phased it out. In addition, Blue Plains was also able to reduce the need for certain emergency planning efforts and regulatory paperwork.[36]

4.1.1.1.1 From Chlorine Gas to Liquid Bleach

In recent years, at least 166 water utilities have reported switching from chlorine gas to liquid bleach, noting that liquid chlorine bleach is safer to work with than chlorine gas.[29] Chemical costs tend to be higher for liquid bleach than chlorine gas, but overall costs are competitive when the full dangers and costs of safety and security are considered, according to plant managers.

More than 33 million people are no longer at risk of being exposed to toxic gas from these water utilities. Hazards remain at the few facilities that manufacture the liquid bleach. Nonetheless, shipping chlorine gas to many locations is arguably more hazardous than securing a few manufacturing facilities in less populated areas. Other substitutes for chlorine gas, such as ultraviolet light or dilute bleach generated on-site, do not involve off-site chemical manufacturing and bulk storage.

In New Jersey alone, hundreds of water treatment plants have stopped using or reduced their use of chlorine gas to below threshold levels as a result of the state's Toxic Catastrophe Prevention Act from 575 such water treatment facilities in 1988 to just 22 in 2001.[38]

Concrete examples of facilities that switched from chlorine gas to liquid bleach, and estimates of the number of people affected:

Utility or Facility	Location	Population no Longer at Risk
City of Wilmington Water Pollution Control	Wilmington, DE	560,000 people
Middlesex County Utilities Authority	Sayreville, NJ	10.7 million people
Metropolitan Wastewater Treatment Plant	St. Paul, MN	520,000 people
Nottingham Water Treatment Plant	Cleveland, OH	1.1 million people
Blue Plains Wastewater Treatment Plant	Washington, D.C.	1.7 million people

4.1.1.1.2 From Chlorine Gas to Ultraviolet Light

The CAP study identified 42 facilities that switched from chlorine gas to ultraviolet light for water treatment, eliminating chemical danger to over 3.5 million people. The use of ultraviolet light also eliminates the related hazards of transporting and working with chlorine gas.

More than 3000 water facilities in the United States now use ultraviolet light, primarily in wastewater treatment, and its functional and economic viability is proven. More drinking water facilities are expected to use ultraviolet light, often in conjunction with other treatments, as a result of new EPA regulations to reduce disinfection byproducts and enhance surface water treatment.[39]

Ultraviolet light and other options, such as ozone, are actually more effective than chlorine against certain biological agents, such as anthrax, that could contaminate drinking water. A multiple barriers approach, such as ultraviolet light and bleach with appropriate site security, has the best chance of preventing deliberate contamination of drinking water.

Concrete examples of facilities that switched from chlorine gas to UV light, and estimates of the number of people affected:

Utility or Facility	Location	Population no longer at risk
White Slough Water Pollution Control Facility	Lodi, CA	606,500 people
South Valley Water Reclamation Facility	West Jordan, UT	131,968 people
R. M. Clayton WRC	Atlanta, GA	1.1 million people
Stamford Water Pollution Control Facility	Stamford, CT	70,000 people
Wyandotte Wastewater Treatment Facility	Wyandotte, MI	1.1 million people

4.1.1.1.3 From Chlorine Gas to Bleach Generated On-Site

The Center for American Progress (CAP) study found a dozen facilities that now treat water by generating bleach disinfectant on-site. This practice eliminates bulk storage and transportation of hazardous chemicals. The process uses salt, water, and electricity to produce a dilute bleach solution. Survey respondents noted that this dilute solution is even safer than the stronger bleach that many utilities receive by truck or rail.

Generating bleach on-site virtually eliminates potential community and workplace exposure to toxic chemicals. An estimated 2000 municipal drinking water systems now generate bleach on-site, with additional applications in wastewater, cooling towers, and food processing.

Concrete examples of facilities that switched from chlorine gas to bleach made on-site, and estimates of the number of people affected:

Utility or Facility	Location	Population no Longer at Risk
Ketchikan Chlorination Plant	Ketchikan, AK	5,510 people
Yorba Linda Water District	Placentia, CA	27,000 people
La Vergne Water Treatment Plant	La Vergne, TN	3,400 people
East & West Site Water & Wastewater Facilities	Margate, FL	98,000 people
Edison Filtration Plant and Well Field	South Bend, IN	18,815 people

4.1.1.1.4 Calcium Hypochlorite Solids as Alternative to Chlorine Gas

One wastewater facility, Town of Garner WWTP, Garner, NC, reported switching from chlorine gas to calcium hypochlorite, a solid. This land-disposal facility spray-irrigates some 300 acres of hay fields with over one million gallons of treated wastewater each day. Calcium hypochlorite is less potentially harmful to soil than alternative sodium hypochlorite. Switching to calcium hypochlorite eliminates the risk of a chlorine gas leak to employees and 205 nearby residents.

4.1.2 Chlorine in Manufacturing

4.1.2.1.1 Safer Delivery Method for Gaseous Chlorine
PVS Technologies, in Augusta, GA, manufactures ferric chloride, which is used in the water and wastewater treatment industries as a flocculent and coagulant. The manufacturing process uses chlorine gas, formerly delivered in 90-ton rail cars. The company eliminated rail cars from the site by constructing a direct pipeline to the chlorine producer, a nearby facility. Eliminating rail transportation removes the dangers of filling, moving, and unloading a large vessel, including both likely incidents, such as transfer-hose failures, and potential worst-cases, such as rupture into an area encompassing 290,000 people.

4.1.2.1.2 Alternatives to Gaseous Chlorine in Paper Processing
SCA Tissue (formerly Wisconsin Tissue Mills), in Menasha, WI, is a large recycled paper mill that formerly used chlorine gas as a bleaching aid. The facility revamped the de-inking process to use sodium hydrosulfite and hydrogen peroxide. This change significantly reduced workplace and community chemical hazards, while avoiding costs of complying with pollution rules, such as certain testing, sampling, and permit reporting. Switching to different chemicals eliminated the danger of a chemical release to any of 210,000 people living within the facility's former vulnerability zone.

Wausau-Mosinee Paper Corporation, in Brokaw, WI, manufactures printing and writing paper. The mill switched from chlorine for bleaching pulp to an oxygen and hydrogen peroxide process. This change improved environmental security and safety by eliminating both the danger of a chlorine gas release and chlorine byproducts from waste streams. The change eliminated a chlorine gas vulnerability to an area containing 59,000 people.

Katahdin Paper (formerly Great Northern Paper) in East Millinocket, ME, manufactures newsprint and telephone directory paper. Under new ownership, the mill eliminated chlorine gas and switched to chlorine bleach for treating incoming process water. The change eliminated a vulnerability zone encompassing 3200 nearby residents.

4.1.2.1.3 Bio-Based Bleaching of Paper Pulp
Research studies undertaken at the Georgia Institute of Technology have utilized the catalytic oxidative properties of laccase, an

oxoreductase enzyme found in several natural systems, to improve the physical properties of lignocellulosic pulps in an enhanced environmentally green manner. This work has identified several novel reactions, including: the unique chemical reactivity of laccase-mediated systems with lignin; the ability of laccase to be used as an oxidative biobleaching system for recycled fiber—a previously unrecognized benefit as a pretreatment for kraft pulping technologies; and as a surface activation technology that yields pulp fibers with substantially improved physical properties. The benefits of these discoveries are anticipated to yield useful methods of eliminating hazardous chlorinated chemical wastes, enhanced usage of recycled paper, improved pulping/bleaching efficiencies, thereby reducing the need for virgin wood resources, and improved physical paper properties, thereby reducing the power consumption associated with the production of high-value paper. By virtue of a reduced reliance on chlorine bleaching, the laccase-based bleaching technology also promises to make pulp and paper processing facilities less susceptible to terrorist attack.

4.1.2.1.4 Alternatives to Chlorine Gas in Circuit Board Manufacturing
Photocircuits Corporation, in Glen Cove, NY, manufactures printed circuit boards for use in computers, cars, phones, and many other products. The facility formerly used chlorine gas in the copper etching process used to make circuit boards, but switched to sodium chlorate. This change reduced hazards to employees and eliminated an off-site vulnerability zone that encompassed 21,000 people.

Sanmina-SCI (formerly Hadco), in Phoenix, AZ, manufactures high-end printed circuit boards, and switched from chlorine gas to sodium chlorate in a closed loop system that directly feeds the etching process. The change eliminated the threat of a gas release to employees and 4000 Phoenix residents.

4.1.2.1.5 Alternatives to Chlorine Gas in Metals Processing
Kaiser Aluminum's Trentwood Works, in Spokane, WA, is a large aluminum rolling mill. The facility formerly used large volumes of chlorine gas from 90-ton rail cars in fluxing operations that remove impurities from molten aluminum. Plant managers and workers on the plant's health and safety committee became concerned with recurring chlorine leaks, injuries, and corrosion of tools and infrastructure. After further investigation, the facility changed the fluxing process to a solid magnesium chloride salt injected with nitrogen gas. This change

greatly improved worker safety, reduced maintenance costs, and eliminated the danger of a major chlorine gas release to any of 137,000 nearby residents.

4.1.2.2 Chlorine-Free Synthesis of 4-Aminodiphenylamine

A critically important reaction used to manufacture a wide range of chemical products is nucleophilic aromatic substitution. Unfortunately, this reaction generates a large amount of toxic waste associated with synthesis of both intermediates and products. Of special concern are chlorinated species, the large-scale chemical synthesis of which has come under intense scrutiny. Solutia, Inc. (formerly Monsanto Chemical Company), one of the world's largest producers of chlorinated aromatics, funded research to explore alternative synthetic reactions for manufacturing processes that do not require the use of chlorine and that represent new atom-efficient chemical reactions. Monsanto's Rubber Chemicals Division (now Flexsys America L.P.) has developed a new method for manufacturing a rubber preservative that eliminates chlorine waste at the source.

The research began as an exploration of new routes to a variety of aromatic amines that would not rely on the use of halogenated intermediates or reagents. Of particular interest was the identification of novel synthetic strategies to 4-aminodiphenylamine (4-ADPA), a key intermediate in the Rubber Chemicals family of antidegradants. The total world volume of antidegradants based on 4-ADPA and related materials is approximately 300 million pounds per year, of which Flexsys is the world's largest producer.

The conventional process to 4-ADPA is based on the chlorination of benzene. Since none of the chlorine used in the process ultimately resides in the final product, the ratio of pounds of waste generated to pound of product produced is highly unfavorable. A significant portion of the waste is in the form of an aqueous stream that contains high levels of inorganic salts contaminated with organics that are difficult and expensive to treat. Furthermore, the process also requires the storage and handling of large quantities of chlorine gas.

Flexsys found a solution to this problem in a class of reactions known as nucleophilic aromatic substitution of hydrogen (NASH). Through a series of experiments designed to probe the mechanism of NASH reactions, Flexsys achieved a breakthrough in understanding this chemistry that has led to the development of a new process to 4-ADPA that utilizes the base-promoted, direct coupling of aniline and nitrobenzene.

The discovery of the new route to 4-ADPA and the elucidation of the mechanism of the reaction between aniline and nitrobenzene have been recognized throughout the scientific community as a breakthrough in the area of nucleophilic aromatic substitution chemistry, and hailed as a discovery of a new chemical reaction that can be implemented into innovative and environmentally safe chemical processes.

The environmental benefits of the new coupling process are significant, and include a dramatic reduction in waste generated: In comparison to the process traditionally used to synthesize 4-ADPA, the Flexsys process generates 74% less organic waste, 99% less inorganic waste, and 97% less wastewater.

In global terms, if just 30% of the world's capacity to produce 4-ADPA and related materials were converted to the Flexsys process, 74 million pounds less chemical waste would be generated per year and 1.4 billion pounds less wastewater would be generated per year.

By achieving the green chemistry principle of eliminating waste by simply not creating it at the source, this new chlorine-free process for the production of 4-ADPA also eliminates the possibility that large volumes of chlorine gas and hazardous waste products could be targeted for nefarious purposes.

4.1.2.3 PVC-Free Backing for Carpet Tile

Shaw Industries has developed a new backing for carpet tile called EcoWorx™ that replaces conventional carpet tile backings containing bitumen, polyvinyl chloride (PVC), or polyurethane. The new backing replaces PVC resins with polyolefin resins, which have low toxicity, superior adhesion, dimensional stability, and easy recycling methods. EcoWorx™ carpet tile backing can be readily separated from any type of carpet fiber, allowing the fiber and backing to be recycled separately.

Historically, carpet tile backings have been manufactured using bitumen, polyvinyl chloride (PVC), or polyurethane (PU). While these backing systems have performed satisfactorily, there are several inherently negative attributes due to their feedstocks or their ability to be recycled. PVC holds the largest market share of carpet tile backing systems, and Shaw resolved to design around PVC due to the multitude of health and environmental concerns around vinyl chloride monomer, chlorine-based products, plasticized PVC-containing phthalate esters, and toxic byproducts of combustion of PVC, such as dioxin and hydrochloric acid.

Alternatives to PVC backings already on the market also present drawbacks. Due to the thermoset cross-linking of polyurethanes, they are extremely difficult to recycle, and are typically downcycled or landfilled at the end of their useful life. Bitumen provides some advantages in recycling, but the modified bitumen backings offered in Europe have failed to gain market acceptance in the United States, and are unlikely to do so.

Shaw selected a combination of polyolefin resins from Dow Chemical as the base polymer of choice for EcoWorx™ due to the low toxicity of its feedstocks, superior adhesion properties, dimensional stability, and its ability to be recycled. The EcoWorx™ compound also had to be designed to be compatible with nylon carpet fiber. Although EcoWorx™ may be recovered from any fiber type, compatibility with nylon-6 provides a significant advantage. Polyolefins in EcoWorx™ are compatible with known nylon-6 depolymerization methods, whereas PVC interferes with those processes. Nylon-6 chemistry is well known and not addressed in first-generation production.

From its inception, EcoWorx™ met all of the design criteria necessary to satisfy the needs of the marketplace from a performance, health, and environmental standpoint. Research indicated that separation of the fiber and backing through elutriation, grinding, and air separation proved to be the best way to recover the face and backing components, but an infrastructure for returning postconsumer EcoWorx™ to the elutriation process was necessary. Research also indicated that the postconsumer carpet tile had a positive economic value at the end of its useful life. The cost of collection, transportation, elutriation, and return to the respective nylon and EcoWorx™ manufacturing processes is less than the cost of using virgin raw materials, and Shaw guarantees that it will reclaim EcoWorx™ products at the end of its useful life free of charge to the customer.

With introduction in 1999 and an anticipated lifetime of 10−15 years on the floor, significant quantities of EcoWorx™ began to flow back to Shaw in approximately 2007. An expandable elutriation unit is now operating at Shaw, typically dealing with industrial EcoWorx™ waste. Recovered EcoWorx™ is flowing back to the backing extrusion unit, and EcoWorx™ now contains 40% recycled content. Caprolactam recovered from the elutriated nylon-6 is flowing back into nylon compounding. Due to the success of EcoWorx™, Shaw ceased PVC production in 2004.

By designing around the use of PVC backing, this innovative carpet tile adhesive not only avoids production of unhealthy and unsafe compounds, like vinyl chloride monomer, chlorine-based products, plasticized PVC-containing phthalate esters, and toxic byproducts of combustion of PVC, such as dioxin and hydrochloric acid, but also makes it impossible for these compounds to be released into the environment as a result of a terrorist-provoked catastrophe, or unintended accident at their place of manufacture, transport, storage, or use.

4.1.2.4 Novel and Versatile TAML Oxidant Activators Replace Chlorine Oxidants

A new species of compounds act as activators that catalyze the oxidizing power of hydrogen peroxide. These tetraamido-macrocyclic ligand (TAML™) activators are iron-based, and contain no toxic functional groups. TAML™ activators constitute a significant technology breakthrough for the pulp and paper industry and the laundry field because they are chlorine-free, and work by catalyzing the naturally occurring oxidant, hydrogen peroxide, to prepare wood pulp for papermaking, to remove stains from laundry, and to eliminate dye transfer between clothes during laundry. By obviating the need for chlorine-based oxidants, this new environmentally benign oxidation technology improves safety, eliminates chlorinated organics from wastewater streams, and saves energy and water.

In nature, selective oxidation is achieved through complex mechanisms keyed to a limited set of elements available and/or plentiful in the environment. In the laboratory, chemists favor a simpler design that utilizes the full range of the periodic table; however, some of these elements cause lasting environmental damage. The problem of persistent pollutants in the environment can be minimized by employing reagents and processes that mimic those found in nature. By developing a series of activators effective with the natural oxidant, hydrogen peroxide, Professor Terry Collins of Carnegie Mellon University has devised an environmentally sound oxidation technique with widespread applications.

The key to quality papermaking is the selective removal of lignin from the white, fibrous polysaccharides, cellulose, and hemicellulose. Wood-pulp delignification has traditionally relied on chlorine-based processes that produce chlorinated pollutants. Collins has demonstrated that TAML™ activators effectively catalyze hydrogen peroxide in the selective delignification of wood pulp. This is the first low-temperature peroxide oxidation technique for treating wood pulp, signifying valuable energy savings for the industry. Environmental compliance costs can also be expected to decrease with this new approach because chlorinated organics are not generated in this totally chlorine-free process.

TAML™ activators may also be applied to the laundry field, where most bleach products are based on peroxide. When bound to fabric, most commercial dyes are unaffected by the TAML™-activated peroxide. However, random molecules of dye that "escape" the fabric during laundering are intercepted and destroyed by the activated peroxide before the dye has a chance to transfer to other articles of clothing. This technology prevents dye-transfer accidents while offering improved stain-removal capabilities. Washing machines that require less water will be practical when the possibility of dye-transfer is eliminated.

Another active area of investigation is the use of TAML™ peroxide activators for water disinfection. Ideally, the activators would first kill pathogens in the water sample, then destroy themselves in the presence of a small excess of peroxide. This protocol could have global applications, from developing nations to individual households.

The versatility of the activators in catalyzing peroxide has been demonstrated in the pulp and paper and laundry industries. Environmental benefits include decreased energy requirements, elimination of chlorinated organics from the waste stream, and decreased water use. The development of new activators and new technologies will provide environmental advantages in future applications.

Because TAML™ activators convert the relatively innocuous hydrogen peroxide into a much more efficient and potent oxidant, there will be fewer instances, where chlorine-based agents are required. In turn, this will reduce the chances that hazardous quantities of chlorine gas, liquids, or chlorinated waste could pose a threat to public health and safety.

4.2 HYDROGEN CYANIDE

Background: Hydrogen cyanide (HCN) is a highly poisonous, colorless and volatile liquid that turns to gas at 26 °C. Cyanides, which are the salts of hydrogen cyanide, are used in dyeing, acrylic plastics, tempering steel, explosives, and engraving. The largest industrial applications are for etching and finishing surfaces, and in the extraction of gold and other precious metals. As a product of combustion, it is found in vehicle exhaust, burning tobacco, burning plastics, and also the pits of certain fruits, such as cherries, apricots, and bitter almonds.

Though theoretically detectable as a faint bitter odor, many people cannot smell it due to a genetic trait. Hydrogen cyanide has been the cause of accidental poisonings in chemistry labs when acid combines with cyanides to form the gaseous HCN. Exposure to lower levels may result in breathing difficulties, heart pains, vomiting, blood changes, headaches, and enlargement of the thyroid gland. High concentrations cause brain and heart damages, and may lead to coma or death.[40]

Employed by Nazi Germany for mass executions during World War II, hydrogen cyanide gas has potential to cause great human loss if wielded as a chemical weapon. U.S. intelligence agents now believe that al-Qaeda operatives intended to use hydrogen cyanide gas for attacks planned in the New York subway system,[41] and some speculate for the foiled terrorist plot in London in August, 2006.[42]

Green chemistry methods that help to reduce the total amount of cyanide and/or hydrogen cyanide produced, used, stored, and transported will concomitantly reduce vulnerability to terrorist attacks targeting the chemical sector. What follows are descriptions of some cases, where products or chemicals that conventionally require HCN in their synthesis have successfully reduced or replaced that reliance on HCN in favor of less hazardous alternatives.

4.2.1.1.1 Synthesis of Iminodisuccinate, a Biodegradable Chelating Agent

Chelating agents are used in a variety of applications, including detergents, agricultural nutrients, and household and industrial cleaners. Most traditional chelating agents do not break down readily in the environment, and require large volumes of hydrogen cyanide and formaldehyde in their synthesis. Bayer Corporation and Bayer AG have developed a 100% waste-free, environmentally friendly manufacturing process for a new molecule that acts as a chelating agent, and is readily biodegradable and nontoxic, called D,L-aspartic-N-(1,2-dicarboxyethyl) tetrasodium salt, also known as sodium iminodisuccinate. The process for synthesizing this new molecule eliminates the use of formaldehyde and hydrogen cyanide.

Sodium iminodisuccinate belongs to the aminocarboxylate class of chelating agents. Nearly all aminocarboxylates in use today are acetic acid derivatives produced from amines, formaldehyde, sodium hydroxide, and hydrogen cyanide. The industrial use of thousands of tons of hydrogen cyanide is an extreme toxicity hazard. In contrast, sodium iminodisuccinate is produced from maleic anhydride (a raw material also produced by Bayer), water, sodium hydroxide, and ammonia. The only solvent used in the production process is water, and the only side product formed—ammonia dissolved in water—is recycled back into sodium iminodisuccinate production or used in other Bayer processes.

Most traditional chelating agents, however are poorly biodegradable. Some are actually quite persistent and do not adsorb at the surface of soils in the environment or at activated sludge in wastewater treatment plants. Because of this poor biodegradability combined with high water solubility, traditionally used chelators are readily released into the environment, and have been detected in the surface waters of rivers and lakes, and in make-up water processed for drinking water.

Sodium iminodisuccinate is characterized by excellent chelation capabilities, especially for iron(III), copper(II), and calcium, and is both readily biodegradable and benign from a toxicological and ecotoxicological standpoint.

Sodium iminodisuccinate can be used in a variety of applications that employ chelating agents, for example, as a builder and bleach stabilizer in laundry and dishwashing detergents to extend and improve the cleaning properties of the 8 billion pounds of these products that are used annually. Specifically, sodium iminodisuccinate chelates calcium to soften water and improve the cleaning function of the surfactant. In photographic film processing, sodium iminodisuccinate complexes metal ions, and helps to eliminate precipitation onto the film surface. In agriculture, chelated metal ions help to prevent, correct, and minimize crop mineral deficiencies. Using sodium iminodisuccinate as the chelating agent in agricultural applications eliminates the problem of environmental persistence common with other synthetic chelating agents.

Because Bayer's sodium iminodisuccinate chelating agent eliminates the need for HCN and formaldehyde in chelator production, it not only offers the environmental benefits of biodegradability and waste-free manufacturing, but also helps eliminate all vulnerability to terrorist attack associated with manufacturing, storing, and transporting large volumes of those toxic inputs.

4.2.1.1.2 Green Biocatalysts for Production of Atorvastatin
Codexis, Inc. has developed an enzyme-based process that has greatly improved the yield, efficiency, and safety record for manufacture of atorvastatin, the key building block for Lipitor®, one of the world's best-selling drugs, that lowers cholesterol by blocking its synthesis in the liver. The new enzymatic process is dramatically faster and more efficient than previous methods, and also reduces cyanide-related waste, the use of solvents, and the need for purification equipment.

Atorvastatin calcium is the active ingredient of Lipitor®. Lipitor® is the first drug in the world with annual sales exceeding $10 billion. The key chiral building block in the synthesis of atorvastatin is ethyl (R)-4-cyano-3-hydroxybutyrate, known as hydroxynitrile (HN). Annual demand for HN is estimated to be about 440,000 pounds. Traditional commercial processes for HN require a resolution step with 50% maximum yield or syntheses from chiral pool precursors; they also require hydrogen bromide to generate a bromohydrin for cyanation. All previous commercial HN processes ultimately substitute cyanide for halide under heated alkaline conditions, forming extensive byproducts. They require a difficult high-vacuum fractional distillation to purify the final product, which decreases the yield even further.

Codexis designed a green HN process around the exquisite selectivity of enzymes and their ability to catalyze reactions under mild, neutral conditions to yield high-quality products. Codexis developed each of three enzymes using state-of-the-art, recombinant-based, directed evolution technologies to provide the activity, selectivity, and stability required for a practical and economic process. The evolved enzymes are so active and stable that Codexis can recover high-quality product by extracting the reaction mixture. In the first step, two evolved enzymes catalyze the enantioselective reduction of a prochiral chloroketone (ethyl 4-chloroacetoacetate) by glucose to form an enantiopure chlorohydrin. In the second step, a third evolved enzyme catalyzes the novel biocatalytic cyanation of the chlorohydrin to the cyanohydrin under neutral conditions (aqueous, pH $\sim$7, 77–104 °F, atmospheric pressure). On a biocatalyst basis, the evolved enzymes have improved the volumetric productivity of the reduction reaction by approximately 100-fold, and that of the cyanation reaction by approximately 4000-fold. The process involves fewer unit operations than earlier processes, most notably obviating the fractional distillation of the product.

The process provides environmental and human health benefits by increasing yield, reducing the formation of byproducts, reducing the generation of waste, avoiding hydrogen gas, reducing the need for solvents, reducing the use of purification equipment, and increasing worker safety. By providing a way to produce the product while avoiding the generation of hazardous cyanide waste and use of solvents, the method also helps avoid any situation where the plant might be made the target of an attack.

4.2.1.1.3 Catalytic Dehydrogenation of Diethanolamine
Monsanto has developed a new method for production of the herbicide that eliminates most manufacturing hazards and all use of cyanide, ammonia, and formaldehyde in the synthesis of the key intermediate, DSIDA. Monsanto's novel catalytic synthesis of DSIDA uses a copper catalyst, and is safer because it is endothermic, produces higher overall yield, and has fewer process steps.

Disodium iminodiacetate (DSIDA) is a key intermediate in the production of Monsanto's Roundup® herbicide, an environmentally friendly, nonselective herbicide. Traditionally, Monsanto and others have manufactured DSIDA using the Strecker process requiring ammonia, formaldehyde, hydrochloric acid, and hydrogen cyanide. Hydrogen cyanide is acutely toxic and requires special handling to minimize risk to workers, the community, and the environment. Furthermore, the chemistry involves the exothermic generation of potentially unstable intermediates, and special care must be taken to preclude the possibility of a runaway reaction. The overall process also generates up to one pound of waste for every seven pounds of product, and this waste must be treated prior to safe disposal.

Monsanto has developed and implemented an alternative DSIDA process that relies on the copper-catalyzed dehydrogenation of diethanolamine. The raw materials have low volatility and are less toxic. Process operation is inherently safer, because the dehydrogenation reaction is endothermic and, therefore, does not present the danger of a runaway reaction. Moreover, this zero-waste route to DSIDA produces a product stream that, after filtration of the catalyst, is of such high quality that no purification or waste cut is necessary for subsequent use in the manufacture of Roundup®. The new technology represents a major breakthrough in the production of DSIDA, because it avoids the use of cyanide and formaldehyde, is safer to operate, produces higher overall yield, and has fewer process steps.

The metal-catalyzed conversion of amino-alcohols to amino acid salts has been known since 1945. Commercial application, however, was not known until Monsanto developed a series of proprietary catalysts that made the chemistry commercially feasible. Monsanto's patented improvements on metallic copper catalysts afford an active, easily recoverable, highly selective, and physically durable catalyst that has proven itself in large-scale use.

This catalysis technology can also be used in the production of other amino acids, such as glycine. It is also a general method for conversion of primary alcohols to carboxylic acid salts, and is potentially applicable to the preparation of many other agricultural, commodity, specialty, and pharmaceutical chemicals.

By eliminating the need for hydrogen cyanide, ammonia, formaldehyde, and hydrochloric acid, this new method for production of

RoundUp® also eliminates the possibility of a catastrophic accident caused by would-be terrorists targeting those chemicals in an attack on Monsanto's production facility.

4.2.1.2 HCN and Synthesis of Methyl Methacrylate

1. *Conventional synthesis of methyl methacrylate using HCN*

   Synthesis of methyl methacrylate is fundamental to the production of the transparent plastic polymethyl methacrylate (PMMA), and is estimated at over two million metric tons per year. The monomer is most commonly synthesized via the well-established Acetone Cyanohydrin (ACN) process, as shown below, based on easily available raw materials such as, acetone, hydrogen cyanide, methanol and sulfuric acid. Reaction of acetone and hydrogen cyanide yields acetone cyanohydrin as an intermediate, which is then reacted with excess amount of concentrated sulfuric acid, followed by thermal cracking to form methacrylamide sulfate. The methacrylamide sulfate intermediate is then further hydrolyzed and esterified with aqueous methanol to form methyl methacrylate.

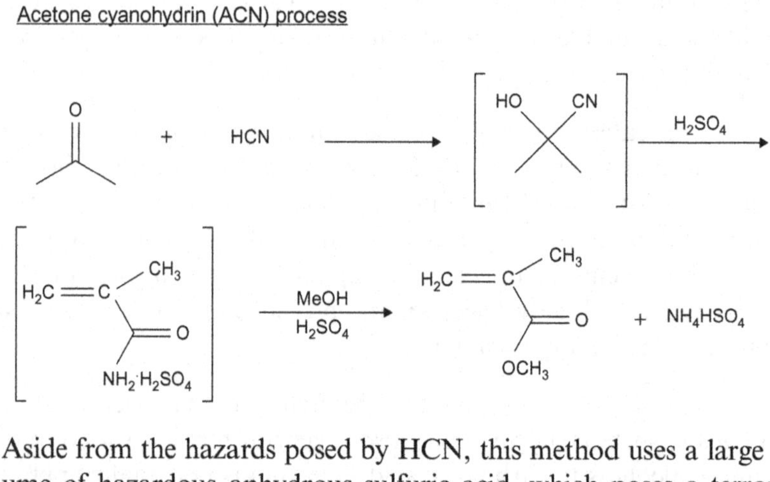

Acetone cyanohydrin (ACN) process

Aside from the hazards posed by HCN, this method uses a large volume of hazardous anhydrous sulfuric acid, which poses a terrorism risk (see sections on sulfuric acid and sulfur dioxide). It also generates a large volume of acidic ammonium bisulfate (NH_4HSO_4), typically as much as 2.5 kg of ammonium bisulfate for every kg of product. This waste acid is either treated with ammonia for conversion to fertilizer grade ammonium sulfate, or is burned in an acid recovery plant for conversion to sulfuric acid for recycling.

2. *HCN-free, C₄-Oxidation Process for Synthesis of Methyl Methacrylate (Isobutylene Catalytic Oxidation):*

Mitsubishi Rayon Co. of Japan has developed a two-stage catalytic oxidation technology to produce methyl methacrylate. Isobutylene or t-butyl alcohol is typically the primary feedstock for this type of reaction (see below). In the first stage, the isobutylene is catalytically oxidized to methacrolein. In the second stage, the methacrolein is oxidized to methacrylic acid. After separation from the by-products, the resulting methacrylic acid is esterified with methanol to methyl methacrylate. The catalysts commonly used in the first-stage oxidation are complex metal oxides of molybdenum, bismuth, cobalt, iron, nickel, alkali metal, antimony, tellurium, phosphorous, and tungsten. The second-stage catalysts are heteropolyacids of molybdenum or phosphomolybdates with metal oxides, such as bismuth, antimony, thorium, chromium, copper, vanadium, or zirconium oxides. In this C4-oxidation process, a basic hydrocarbon feedstock (isobutylene or t-butyl alcohol) is used; other required reactants are air, methanol, and the selected catalysts. Since isobutylene is a major component in C4 streams from ethylene manufacturing plants, C4-oxidation technology may become the most probable basis for future methyl methacrylate production.

Isobutylene catalytic oxidation

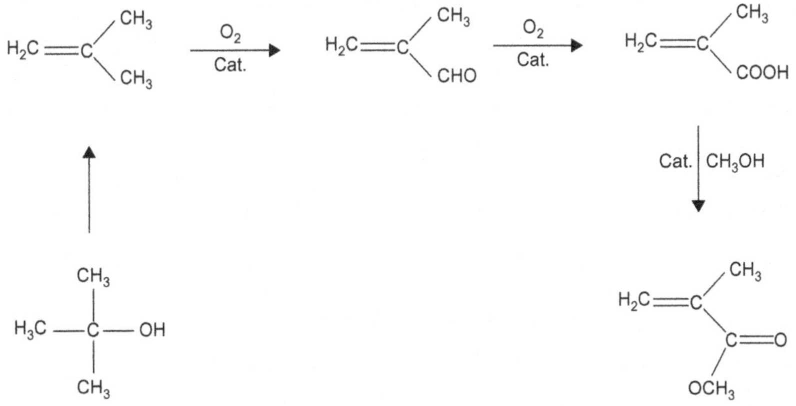

3. *Synthesis of Methyl Methacrylate Via Propionate—Formaldehyde Route:*

Methyl methacrylate or methacrylic acid can be synthesized through the vapor-phase catalytic condensation of either methyl propionate or propionic acid with formaldehyde, as shown below.

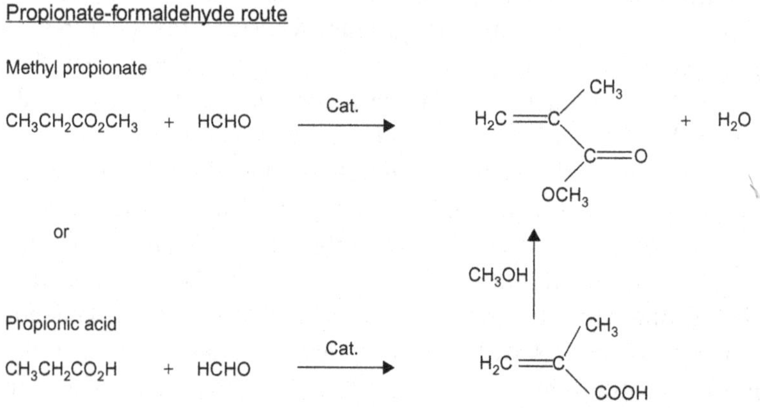

Propionate-formaldehyde route

Methyl propionate

$$CH_3CH_2CO_2CH_3 \ + \ HCHO \ \xrightarrow{\text{Cat.}} \ H_2C=C\begin{smallmatrix} CH_3 \\ C=O \\ OCH_3 \end{smallmatrix} \ + \ H_2O$$

or

$$CH_3OH$$

Propionic acid

$$CH_3CH_2CO_2H \ + \ HCHO \ \xrightarrow{\text{Cat.}} \ H_2C=C\begin{smallmatrix} CH_3 \\ COOH \end{smallmatrix}$$

Methacrylic acid produced from propionic acid in this process can be esterified with methanol to yield methyl methacrylate. Catalysts used in this route include alkali metal or alkaline-earth metal aluminosilicates, potassium hydroxide- or cesium hydroxide-treated pyrogenic silica, alumina, and lanthanum oxide. Both propionic acid and methyl propionate are commercially available through the Oxo process (i.e., by the carbonylation of ethylene or by the hydroformylation of ethylene to propionaldehyde, followed by oxidation of the aldehyde to the corresponding acid). This alternative route has, however, shown only 50% conversion and >80% selectivity rates, and it appears that additional catalyst development is necessary in order to make this process more attractive.

4. *Synthesis of Methyl Methacrylate Via Methyl Propionate Catalytic Oxidation Route:*

Mitsubishi Rayon Co. of Japan also reported a manufacture of methyl methacrylic acid or its methyl ester from a catalytic oxidation of propionic acid or its methyl ester with titanium-vanadium-phosphorous-oxide (PVTiO) as its catalytic system. This reaction is very similar to the propionate-formaldehyde route except that it does not involve use of formaldehyde in the process. The reaction mechanism of this process is still unknown.

Mitsubishi rayon's methyl propionate catalytic oxidation

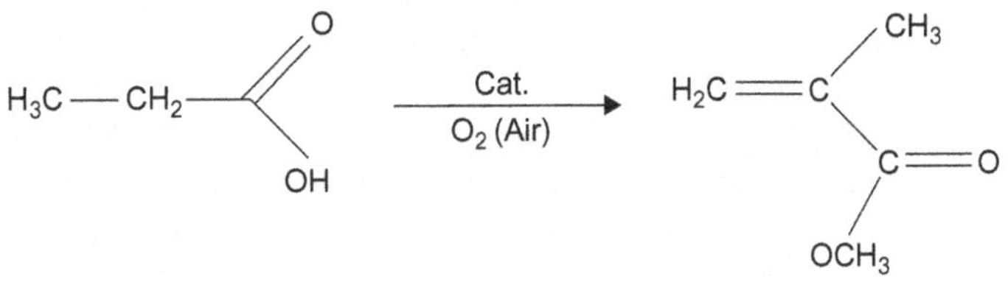

5. *Synthesis of Methyl Methacrylate Via Catalytic Carbonylation of Methylacetylene:*

Shell Research out of Amsterdam reported a novel methyl methacrylate technology by carbonylation of methylacetylene (propyne) with carbon monoxide in the liquid phase in the presence of methanol and catalysts containing palladium compounds, p-toluenesulfonic acid, and diphenyl-2-pyridylphosphine. This process yielded the desired product with 99% selectivity and an average conversion rate of 20,000 mol methyacetylene per gram-atom of palladium per hour. In addition, according to an analysis by SRI International, the process may prove far cheaper than existing manufacturing processes.[43] For example, the Shell technology could make methyl methacrylate for 44 cents per pound in a 100-million lbs/yr plant, compared with other commercial and developmental processes that cost roughly 47–58 cents per pound in a 250-million lbs/yr plant. Disadvantages of this process include the lack of availability and high cost of methylacetylene feedstock. For the process to be commercially successful, the availability and cost concerns of the feedstock must be addressed.

Shell's catalytic carbonylation of methylacetylene

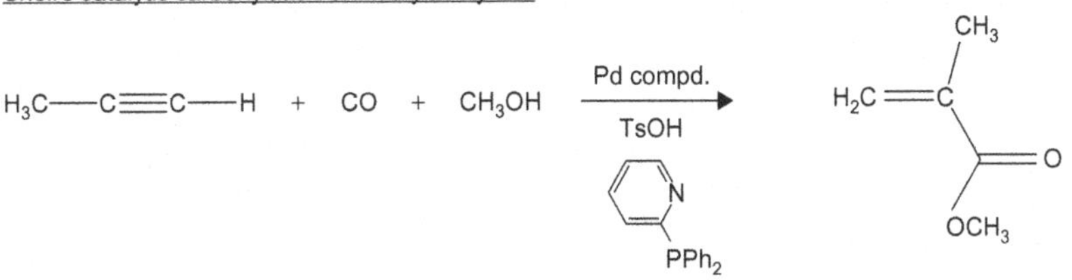

4.2.1.3 HCN in Synthesis of Amino Acids from Aldehydes, Through Catalytic Amidocarbonylation of Syngas (CO/H₂)

Recently, hydrogen cyanide has become a core technology commonly used in the synthesis of amino acids and their derivatives. "Benign-by-design" alternative synthetic pathways exist to avoid or minimize the use of toxic chemicals vulnerable to terrorist attack, such as, hydrogen cyanide, chlorine, and ammonia. Examples identified here highlight alternative synthetic routes for two amino acids and one specialty chemical.

1. *Conventional Synthesis of Amino Acids Using HCN (Strecker Reaction):*
   While most amino acids are obtained from natural sources or via fermentation processes, they also can be chemically synthesized. The conventional route to synthesize amino acids follows the Strecker reaction, as shown in the following reaction sequences. Reaction of hydrogen cyanide with aldehydes together with ammonia yields α-aminonitriles, which can then be hydrolyzed to racemic α-amino acids.

Strecker reaction

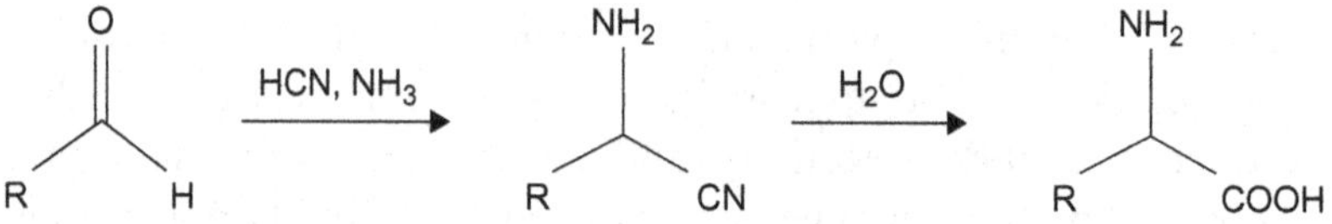

The enantiomerically pure amino acids also can be produced through a similar synthetic pathway catalyzed by enzymes. For example, reaction of hydrogen cyanide with benzaldehyde catalyzed by either (R)- or (S)-oxynitrilase enzyme yields the enantiomeric cyanohydrins (R)- or (S)-mandelonitrile. Alternatively, by adding carbon dioxide and hydrogen cyanide and ammonia as feedstocks, aldehydes can be converted to hydantoins (4-alkylimidazolidine-2,5-diones), which can be then hydrolyzed with either D- or L- hydantoinases to produce D- or L-α-amino acids, respectively.

Enzymatic conversion of hydantoins

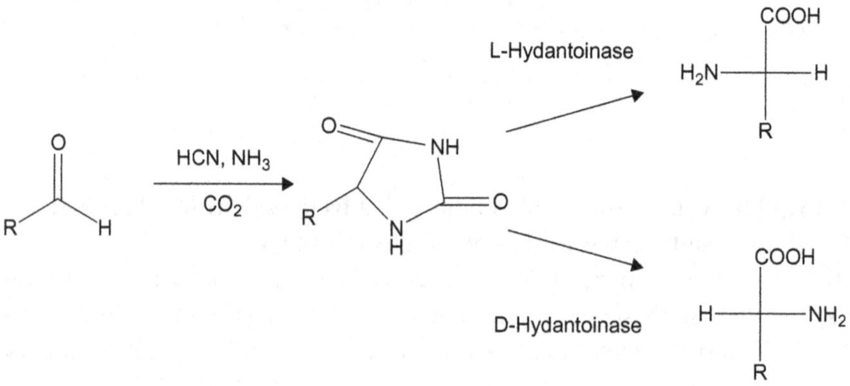

4.2.1.4 Three Alternative HCN-free Syntheses of Amino Acids by Catalytic Amidocarbonylation:

1. *Synthesis of beta-Phenylalanine from amidocarbonylation of phenylacetaldehyde:*

   The amidocarbonylation reaction offers a unique synthetic method to construct two functionalities (amido and carboxylate) simultaneously. This type of reaction uses carbon monoxide and hydrogen (syngas) as building blocks catalyzed by cobalt catalysts and various co-catalysts (e.g., phosphorus, nitrogen, arsenic, sulfur, selenium, or rhodium-containing complexes). For example, β-phenylalanine, a key intermediate in the synthesis of the sweetener aspartame, can be prepared from phenylacetaldehyde by this amidocarbonylation method. Phenylacetaldehyde, acetamide, and syngas (CO/H_2) undergo amidocarbonylation reactions in the presence of a cobalt catalyst ($CO_2(CO)_8$) to yield N-acetyl-β-phenylalanine in 72 mol% yield. Acidic hydrolysis then gives the desired amino acid in high yield with >98% cobalt recovery and acetic acid as a byproduct.

Amidocarbonylation of phenylacetaldehyde:

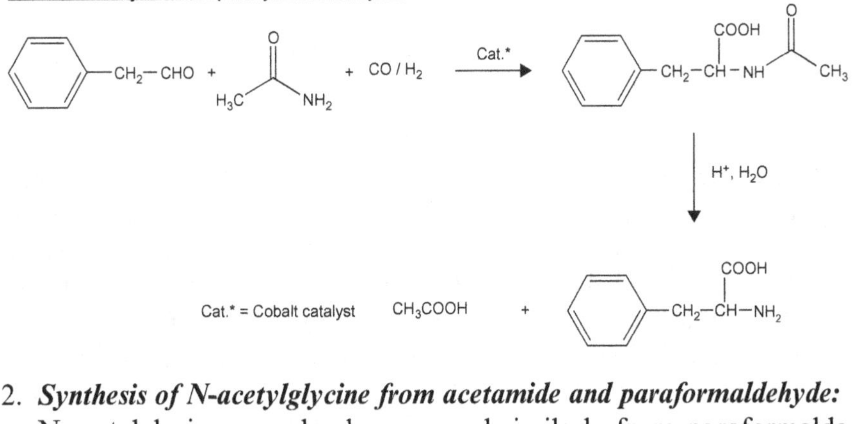

2. *Synthesis of N-acetylglycine from acetamide and paraformaldehyde:*
   N-acetylglycine can also be prepared similarly from paraformaldehyde in good yields (63–78%).

Amidocarbonylation of p-formaldehyde:

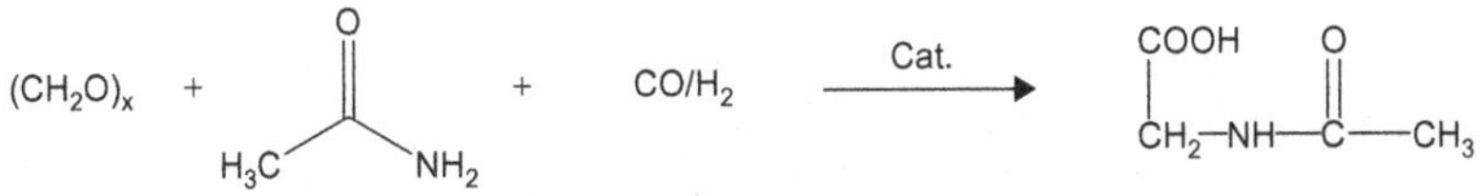

3. *HCN-free synthesis of the specialty chemical, Sarcosinate:*
Sarcosinate, a specialty surfactant, is currently made by acylation of
naturally occurring sarcosine with an acyl chloride, but in reportedly
poor yields.

Sarcosinates via acylation:

$$C_{11}H_{23} - COCl + HOOC - CH_2 - NH - CH_3 \longrightarrow HOOC - CH_2 - NH - \overset{\overset{CH_3}{|}}{\underset{}{}} \overset{C_{11}H_{23}}{\underset{O}{}} + HCl$$

Sarcosine   Sarcosinate

Alternatively, the free carboxylic acid can react with methylamine to
form the corresponding N-methylamide (a secondary amide moiety),
followed by amidocarbonylation with paraformaldehyde and syngas
(CO/H_2) in the presence of dicobalt octacarbonyl $(CO_2(CO)_8)$ at
120 °C to give N-acyl sarcosinate in excellent yields. The product
selectivity is typically 95% at 92% N-methylamide conversion. This
synthesis approach avoids use of acid chloride, and allows the introduction of both the carboxylic acid and secondary amide functionalities in a single step.

Sarcosinates via amidocarbonylation:

$$C_{11}H_{23}\text{-CHOOH} + H_3C - NH_2 \xrightarrow{-H_2O} H_3C - NH_2 \overset{C_{11}H_{23}}{\underset{O}{}} \xrightarrow[(CH_2O)_x]{CO/H_2} HOOC - CH_2 - NH \overset{\overset{CH_3}{|}}{\underset{O}{}} \overset{C_{11}H_{23}}{}$$

Sarcosinate

These three examples demonstrate the applications of amidocarbonylation technology as an alternative route to various amidoacid products, including amino acids and surface active agents (surfactants).
This new amidocarbonylation technology represents a significant ability to make a host of specialty chemicals by using synthesis gas (syngas, consisting primarily of carbon monoxide, carbon dioxide, and
hydrogen) as the basic building block with inexpensive aldehydes independent of natural products. Most importantly, it avoids use of hydrogen cyanide, ammonia, or acid chloride (chlorine) in the synthetic
processes and provides environmentally safer process practices.

4.2.1.4.1 Greener Synthesis of Aprepitant, a Therapy for
Chemotherapy-Induced Discomforts
Aprepitant is the active ingredient in Emend®, a new therapy that—
when used during and shortly after chemotherapy—has been clinically

shown to reduce nausea and vomiting, the most common side effects associated with the chemotherapeutic treatment of cancer.

Aprepitant, which has two heterocyclic rings and three stereogenic centers, is a challenging synthetic target. Merck's first-generation commercial synthesis required six synthetic steps, and was based on the discovery synthesis. The raw material and environmental costs of this route, however, along with operational safety issues compelled Merck to discover, develop, and implement a completely new route to aprepitant.

Merck's new route to aprepitant demonstrates several important green chemistry principles. Whereas the first-generation synthesis required stoichiometric amounts of an expensive, complex chiral acid as a reagent to set the absolute stereochemistry of aprepitant, the new synthesis, in contrast, incorporates a chiral alcohol as a feedstock, and this alcohol is itself synthesized in a catalytic, asymmetric reaction. Merck uses the stereochemistry of this alcohol feedstock in a practical crystallization-induced asymmetric transformation to set the remaining stereogenic centers of the molecule during two subsequent transformations. The new process nearly doubles the yield of the first-generation synthesis. Much of the chemistry developed for the new route is novel, and has wider applications. In particular, the use of a stereogenic center that is an integral part of the final target molecule to set new stereocenters with high selectivity is applicable to the large-scale synthesis of other chiral molecules, especially drug substances.

Implementing the new route has drastically improved the environmental impact of aprepitant production. Merck's new route eliminates all of the operational hazards associated with the first-generation synthesis, including those of sodium cyanide, dimethyl titanocene, and gaseous ammonia. The shorter synthesis and milder reaction conditions have also reduced the energy requirements significantly. Most importantly, the new synthesis requires only 20% of the raw materials and water used by the original one. By adopting this new route, Merck has eliminated approximately 41,000 gallons of waste per 1000 pounds of aprepitant that it produces.

The alternative synthetic pathway for the synthesis of aprepitant is an example of minimizing environmental impact while greatly reducing production costs. Because Merck implemented the new synthesis during its first year of production of Emend®, the company will realize the benefits of this route for virtually the entire lifetime of the product, demonstrating that green chemistry solutions can be aligned with cost-effective ones. Furthermore, by devising an efficient method that requires only 20% of the raw materials and eliminates approximately 41,000 gallons of waste for every 1000 pounds of drug produced, smaller quantities of chemicals are present at the manufacturing facility, and there is less risk of a large-scale accident occurring.

4.3 HCN AND REDUCING VULNERABILITY OF THE CHEMICAL SUPPLY CHAIN

4.3.1.1 On-site, Small Scale Synthesis of Hydrogen Cyanide (HCN)

Hydrogen cyanide is a large volume-hazardous chemical that serves diverse functions. DuPont is the single largest manufacturer of HCN in the world, producing 500 million pounds each year, both for internal use in synthesis and for external sales. The majority of HCN manufactured at DuPont is used internally, primarily for adiponitrile synthesis en route to production of nylon. In that case, HCN is reacted with butadiene to produce adiponitrile, an intermediate in the manufacture of nylon polymer. In addition, over one million lbs per year are shipped out in gas cylinders (approximately 4000 cylinders/year).

On-site synthesis of HCN by the end-use customer would greatly help to minimize health and safety concerns, and would result in a more cost effective synthesis of the derived specialty chemicals. Several options exist for small-scale synthesis of HCN, and individual companies need only choose the most economical and commercially feasible method for their situation. On-site production and consumption of HCN would contribute significantly to a reduction of risk stemming from transport, storage, and handling of HCN.

4.3.1.1.1 Conventional, Large-Scale Synthesis of HCN and
Transport to Site

At the current time, almost all HCN is produced on a large scale using one of two methods: the Andrussow process or the Degussa process (see descriptions below). In addition to nylon manufacture, HCN is used for synthesis of many specialty chemicals. Companies, such as DuPont, currently supply the HCN for these processes via transport of HCN in rail cars from the main large-scale synthesis plants, and repackaging into cylinders, which are then shipped to the end-use customer. Both of these steps (railway and cylinder shipping) create a highly hazardous situation with possible accidental or terrorist release scenarios.

4.3.1.1.2 Alternative Approach: On-site Synthesis
The Andrussow and Degussa processes account for almost all large-scale synthesis, which could also be scaled down for small-scale synthesis on-demand and on-site by end-users, thereby eliminating the risks associated with large-scale storage and transport. There are a number of options for achieving such on-site, on-demand syntheses:

4.3.1.1.3 Scheme 1. Andrussow Process
In the Andrussow process, ammonia and air are fed to a platinum gauze catalyst consisting of many layers of gauze of 1/2-inch total thickness:

$$CH_4 + NH_3 + O_2/N_2 \xrightarrow{\text{Platinum Gauze}} HCN + H_2O$$

Oxygen is combusted with the methane, providing heat for driving the strongly endothermic reaction of methane with ammonia to form HCN. The overall reaction, then, is exothermic. Remaining challenges associated with this process include purification of HCN gas and handling of wastewater.

4.3.1.1.4 Scheme 2. Degussa Process
The second process for large-scale synthesis, also applicable for small-scale use, is the Degussa method whereby methane and ammonia react to form HCN with an external heat source (external methane combustion):

$$CH_4 + NH_3 \xrightarrow{\Delta} HCN + 3H_2$$

In this method, platinum is coated on the inside of alumina tubes, which are then placed in a fired furnace. The product HCN in this method is of a higher concentration than in the Andrussow process, and no wastewater is produced. A disadvantage is the large financial investment needed for the process.

4.3.1.1.5 Scheme 3. Microwave HCN Synthesis

There are some additional alternatives to the Andrussow or Degussa processes, including:

1. Electric direct combination

$$CH_4 + NH_3 + \text{electric heat}$$

2. Acetone cyanohydrin method

$$CH_3C(OH)(CN)CH_3$$

3. Formamide method

$$HCONH_2 + \text{heat}$$

4. Sodium cyanide approach

$$NaCN + H_2SO_4$$

or

5. Monomethylamine method

$$H_2NCH_3 + \text{heat}$$

A novel approach that remains to be developed is use of microwave energy to heat the catalyst for endothermic synthesis of HCN. As in the Degussa process, methane is directly catalytically converted into HCN using activated carbon, platinum, rhodium, or palladium on alumina. Instead of using a conventional external tube furnace heat source, this method involves heating the catalyst to approximately 1000 °C with a microwave generator. HCN yields that are comparable or better than those achieved in the Degussa process were obtained (90 + %) using the platinum on alumina or carbon catalysts; rhodium or palladium were less selective. An advantage of microwave heating is rapid startup and shutdown, as well as the absence of any waste stream or need for ammonia scrubbing. A disadvantage of this method is the expense of the microwave generator and the technological problems of scaling up a commercial process, specifically because microwaves have limited penetration potential, multiple generators are required, and homogeneous catalyst heating is required.

Although some of these alternative approaches to synthesizing HCN have faced economic and technological challenges limiting their commercial application, continued efforts to make them more widely practicable will lower the need to transport large quantities of HCN and, thereby, reduce exposure to the possibility of terrorist attacks on HCN transport and storage infrastructure.

4.4 HYDROGEN FLUORIDE (HYDROFLUORIC ACID)

Background: Hydrogen fluoride (HF) is a corrosive colorless fuming liquid or gas with a strong irritating odor. It is used to make refrigerants, herbicides, pharmaceuticals, high-octane gasoline, aluminum, plastics, electrical components, and fluorescent light bulbs. Sixty percent of the hydrogen fluoride used in manufacturing is for processes to make refrigerants. It is also used for etching glass and metal.

Hydrogen fluoride goes easily and quickly through the skin and into the tissues in the body. There, it damages the cells and causes them to not work properly. The seriousness of poisoning caused by hydrogen fluoride depends on the amount, route, and length of time of exposure, as well as the age and preexisting medical condition of the person exposed. Breathing hydrogen fluoride can burn lung tissue and cause swelling and fluid accumulation in the lungs (pulmonary edema), and may be fatal. Skin contact with hydrogen fluoride may cause severe burns that develop after several hours and form skin ulcers.

In 1987, a pipe on an anhydrous hydrofluoric acid storage tank was accidentally severed by construction workers at the Marathon refinery in Texas City, Texas, releasing gas that formed a dense vapor cloud of hydrofluoric acid that migrated through the surrounding community. More than 1000 people were treated for injuries and about 4000 had to be evacuated.[44]

The Centers for Disease Control and Prevention has warned that hydrogen fluoride could be used as a chemical terrorist agent.

4.4.1.1 Hydrogen Fluoride in Metals Production

Henkel Surface Technologies, in Calhoun, GA, makes industrial coating products for cleaning and treating metal surfaces. The facility formerly used highly concentrated (70 %) hydrofluoric acid. Henkel switched to less concentrated (less than 49%) hydrofluoric acid as a result of a company-wide safety policy. While still hazardous upon contact, less concentrated hydrofluoric acid in an aqueous solution is less volatile, and does not readily form a toxic gas cloud that can drift off-site if released. The change eliminated a vulnerability zone that is home to 300 nearby residents.

The Ford Meter Box Company, Inc., in Pell City, AL, makes water utility equipment, such as clamps and repair sleeves. The company formerly used hydrofluoric acid in a dip tank to clean and make the surface of metal parts less reactive for use in harsh environments underground. The company switched to a process that uses ammonium bifluoride to generate less hazardous hydrofluoric acid solution. This change eliminated a vulnerability zone encompassing 50 people.

4.4.1.2 HF in Gasoline Alkylation Technology
Under the Clean Air Act Amendment regulations and their directives, there has been increasing demand focused on producing "high-octane" gasoline (unleaded fuel) to meet new gasoline reformulation require-ments. Usually, this octane component, also known as gasoline alkylate, is manufactured from isobutane and primary C4-olefins in the petroleum refinery process, using either hydrogen fluoride or sulfuric acid as catalysts. Alternative solid acid catalysts are in the research phase, and may be available shortly.

4.4.1.2.1 Conventional Processes for Gasoline Alkylate Production
There are two existing acid catalysis processes for the production of gasoline alkylate. The first process involves the use of anhydrous hydro-fluoric acid as a catalyst, and the other one uses sulfuric acid. The synthesis of gasoline alkylate is shown below:

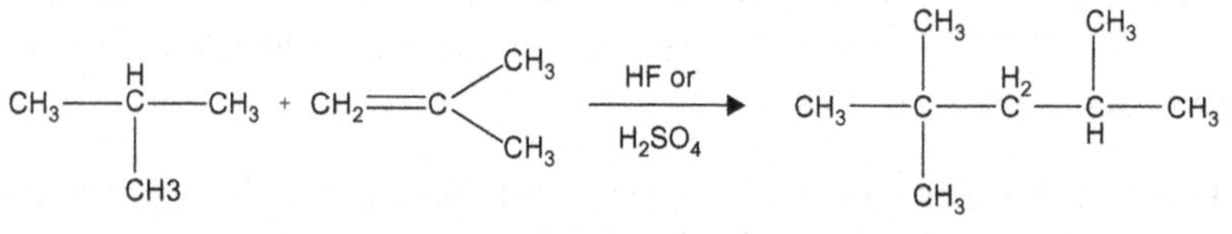

For both processes, the catalytic reaction is carried out in a two-liquid phase system in which it requires a very large volume of liquid acid catalyst. In other words, it requires very large "in-process" inventories of acid catalysts. For example, a typical HF-based process plant that has a 10,000 barrel daily capacity may require as much as 400,000 lbs. of stored anhydrous HF. Petroleum refiners favor the HF-based process since it works well, and all the HF used is recycled within the process. However, HF is highly toxic, and the process is still subject to potentially catastrophic release. Based on environmental concerns and safety issues related to the HF process, much attention has been focused on the reduction of overall in-process HF inventories and the need to mitigate any potential HF releases. It is widely expected that this HF-based technology will be phased out in the near future. For the additional problem regarding the sulfuric acid-based process, please see the sub-section on sulfuric acid.

4.4.1.2.2 Alternative Solid Acid Catalyst to Replace Liquid Acids

Attempts to develop a "solid" catalyst for the replacement of liquid acids have been on-going for years, without much success. For example, a solid catalyst developed based on a silica alumina molecular sieve has not yet led to a commercial process. Potential use of zeolite catalysts as possible solid superacids is still in developmental stages because of the low active-site density of some zeolites and because most of these sites are confined within the zeolite micropores, thus limiting access to smaller molecules and resulting in high catalyst deactivation rates. However, Catalytica Inc. (Mountain View, CA) has joined with Conoco (Stamford, CT) and Neste Oy (Helsinki, Finland) to successfully develop a proprietary technology for a solid acid catalyst and reactor system for the alkylation process[45] (detailed process information from Catalytica is unavailable due to its proprietary status). This new process was tested on a pilot plant scale at Neste's facilities in Porvoo, and the operational test results show this catalyst technology to be as good as or better than current liquid acid catalyst systems. This alternative solid catalyst process is not only environmentally safer, but also offers some advantages, such as production of higher octane value alkylate and reduction of utility consumption by virtue of the new process operating with less feedstock (isobutane) recycling compared with the sulfuric acid process.[46] However, the overall economics of this process still remain a major question, and it continues to be commercialized.

4.4.1.3 Ibuprofen Manufacture and HF

The issues relating to inherently safer chemistry are directly applicable to the pharmaceutical, as well as the industrial chemical sector. In the case of ibuprofen, recent chemical conservation improvements may result in increased worker or population risks via replacement of $AlCl_3$ with HF. Work is underway to replace both $AlCl_3$ and HF with an inherently safer solid recyclable acid catalyst process using zeolites or acid clays.

4.4.1.3.1 Conventional Route for Ibuprofen (Developed by The Boots Company)

Ibuprofen, a widely used over-the-counter analgesic, is made by the classical route developed by the Boots Company as shown below. This process involves six step-wise reactions, including Friedel-Crafts acylation on the first step. Such a process entails too many synthetic steps, thus decreasing the overall yields, and also generating substantial amounts of waste by-products, such as inorganic salts.

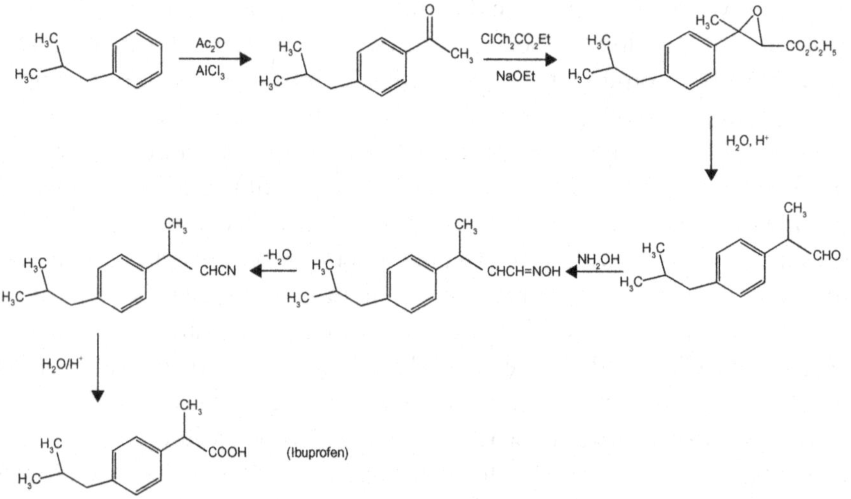

4.4.1.3.2 Alternative Route for Ibuprofen (Developed by Hoechst-Celanese)

An alternative route to make ibuprofen was developed by Hoechst-Celanese (BHC Company) according to the following reaction sequences.

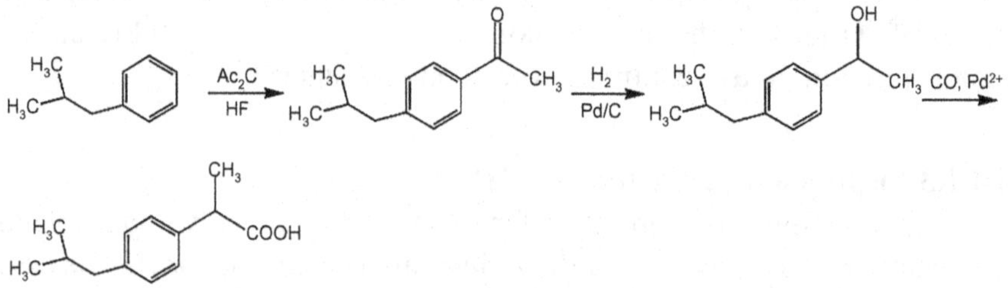

This process involves only three catalytic steps (HF-based acid catalytic acylation, catalytic hydrogenation, and catalytic carbonylation) as opposed to six stoichiometric steps. BHC was the recipient of the

Kirkpatrick Achievement Award for "outstanding advances in chemical engineering technology" in 1993 for the new technology, which boasts approximately 80% atom utilization (virtually 99% including the recovered byproduct acetic acid). The method replaces a previous one comprised of less than 40% atom utilization.

It is interesting to note that both the Boots and Hoechst processes use the same feedstock (isobutylbenzene) and intermediate, p-isobutylacetophenone. The difference between the two lies in the use of an acid catalyst for the acylation step. The Boots process uses the traditional Friedel-Crafts catalyst, aluminum chloride ($AlCl_3$), whereas the alternative Hoechst-Celanese process uses liquid hydrogen fluoride (HF) as a catalyst in place of aluminum chloride. The latter route is more attractive, not only because it uses catalytic technologies to shorten the reaction sequences, but also because there is no salt formations from this process. The use of anhydrous hydrogen fluoride as both catalyst and solvent offers important advantages in reaction selectivity and waste reduction. Large volumes of aqueous wastes (salts) normally associated with such manufacturing are virtually eliminated. The anhydrous hydrogen fluoride catalyst/solvent is recovered and recycled with greater than 99.9% efficiency. No other solvent is needed in the process, simplifying product recovery and minimizing fugitive emissions. The nearly complete atom utilization of this streamlined process makes it an environmentally friendly technology. Commercialized since 1992 in BHC's 7.7-million-pound-per-year facility in Bishop, TX, the new process has been cited as an industry model of environmental excellence in chemical processing technology.

However, to compare alternative routes solely on the basis of waste generation is an oversimplification. An environmentally friendly or favorable process, such as HF-based synthesis of ibuprofen, is not always a "safer" process.

The HF used in this process, although recovered and recycled, remains a highly toxic and corrosive chemical, and requires large on-site inventories. In addition, HF is still subject to potentially catastrophic release. If HF is accidentally released into the air, it forms a potentially lethal cloud that travels at ground level. The safety versus environmentally friendly factors present in the HF-based technologies for manufacture of ibuprofen represent a good example of the factors to be considered when dealing with trade-off issues of alternative synthetic routes.[47]

The uses of clean, catalytic technologies in organic syntheses have gained wide attention, and much effort is being focused on the development of solid recyclable acid catalysts to replace anhydrous aluminum chloride or hydrogen fluoride for Friedel-Crafts acylations. The most promising candidates include zeolites and acidic clays. One type of heterogeneous catalyst proprietarily prepared from impregnation of clay or alumina with typical chemicals under the trade name "envirocats" has been used in various catalytic reactions, such as benzoylations, acetylations, alkyl cations, arenesulfonylations, and air oxidations. These solid acid catalysts offer an alternative for HF or aluminum chloride in regard to a safer chemistry practice.

In the meantime, the HF-based method may not be perfect, but because its catalytic process provides an elegant solution to the prevalent problem of large quantities of solvents and wastes associated with the traditional stoichiometric use of auxiliary chemicals for chemical conversions, it also reduces the likelihood that large volumes of those solvents could be released during a catastrophic attack on the plant.

4.5 PHOSGENE

Background: Though not found naturally in the environment, phosgene ($COCl_2$) is a major industrial chemical used to make many other chemicals, including plastics and pesticides. At room temperature (70 °F), phosgene is a poisonous gas that may appear colorless or as a white to pale yellow cloud. With cooling and pressure, phosgene gas can be converted into a liquid so that it can be shipped and stored; however, when liquid phosgene is released, it quickly turns into a gas that is heavier than air, stays close to the ground, and spreads rapidly. Phosgene can be formed when chlorinated hydrocarbon compounds are exposed to high temperatures.

At low concentrations, phosgene has a pleasant odor of newly mown hay or green corn. It was used extensively during World War I as a choking (pulmonary) agent, and is known in the military as "CG." Among the chemicals used in the war, phosgene was responsible for the large majority of deaths.

Phosgene has been used extensively as an organic synthetic reagent to prepare isocyanates, carbamates, organic carbonates, and chloroformates used in the manufacture of polyurethane and polycarbonate resins, highly toxic and reactive, safety precautions and diligent care are required in process design, plant operation, handling, storage, and transport of this hazardous substance.

Phosgene-based synthesis of carbamate pesticides may pass through the intermediate methyl isocyanate, which was responsible for the catastrophe at the Union Carbide chemical plant in Bhopal, India, in 1984. A relief valve lifted on a storage tank containing methyl isocyanate, releasing a cloud of toxic gas onto residential areas surrounding the plant, ultimately causing the death of more than 15,000 people, and the injury of several hundred thousand more.[22]

In the present day climate moving toward greater protection of health, safety, and the environment, efforts to replace phosgene with safer, greener substitutes in benign-by-design chemical processes are gaining ground. Phosgene itself is commonly produced by the reaction of gaseous carbon monoxide and chlorine. Thus, alternative chemical processes, which involve replacement of phosgene with other feed stocks also ultimately, reduce the demand for reagent chlorine gas. Some examples of new phosgene-free synthetic pathways show potential for replacement of conventional phosgenation processes long used by the chemical industry. It should be noted that there are significant data showing that BPA is an endocrine disruptor, and its inclusion in this section should not be interpreted as an endorsement of its use.

4.5.1.1 Conventional Phosgenation Synthesis of Isophorone Diisocyanate:

Isophorone diisocyanate (IPDI) has traditionally been manufactured from isophorone diamine (IPDA) by the conventional phosgenation route as shown below:

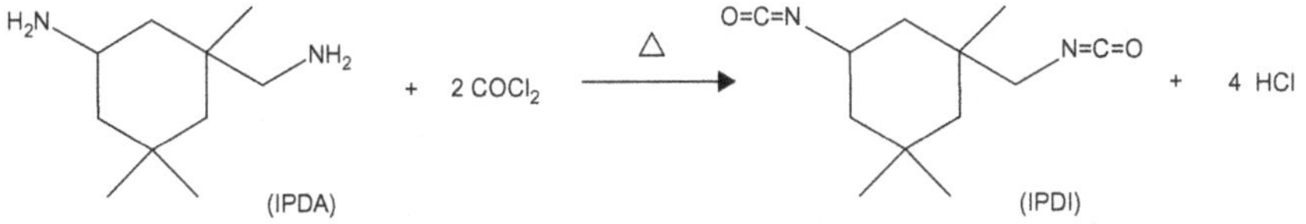

4.5.1.2 Alternative Phosgene-Free Processes

4.5.1.2.1 Urea Adduct-Based, Phosgene-Free Diisocyanate Process Developed by Huls A.G.

Huls in Germany has developed a novel, pilot-scale process for the production of diisocyanates from diamines. This proprietary technology promises to provide another alternative to the conventional phosgenation process by which isophorone diisocyanate (IPDI) is produced for Huls at their Baytown, Texas plant. The new route accomplishes the conversion of isophorone diamine (IPDA) to IPDI using a urea adduct-based process for large scale manufacturing used in the $40 million diisocyanate plant in Theodore, Alabama. It is said to be a highly versatile technology by which virtually any diamine can be converted into a diisocyanate.[48] Other aliphatic diisocyanate raw materials manufacturer by Huls for the coatings industry are said to include the saturated compounds dimethyl diisocyanate and hexamethylene diisocyanate.

4.5.1.2.2 Daicel Phosgene-Free Process to Produce Diisocyanates Via Diurethanes

Daicel Chemical Industries, Ltd., has patented a process for reacting diamines, such as isophorone diamine (IPDA), with dimethyl carbonate.[49] The condensation is accomplished in the presence of an alkaline catalyst, such as methanolic sodium methylate at 70 °C, to produce the diurethane. Treatment of the diurethane with hydrogenated terphenyl containing manganese acetate at 230 °C *in vacuo* produces isophorone diisocyanate by thermal decomposition. This non-phosgene process has a reported product yield of 93% of IPDI containing 1% isophorone monoisocyanate.

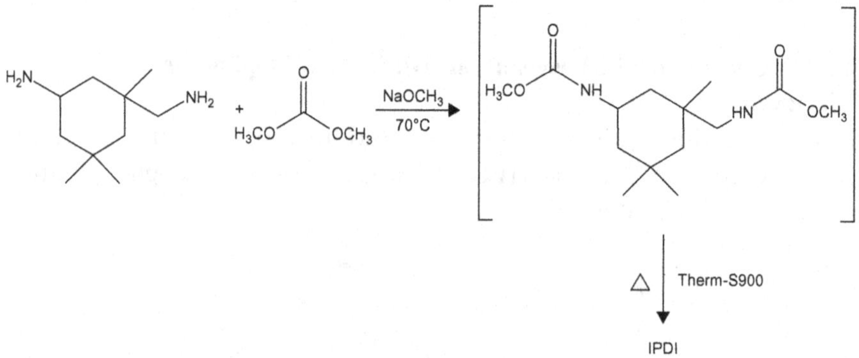

4.5.1.3 Methyl Isocyanate Manufacture, Conventional Method

Conventional route for industrial scale production of isocyanates by phosgenation of corresponding amines (Riley et al., 1994):

$$RNH_2 + COCl_2 \longrightarrow [RNHCOCl] \longrightarrow RNCO + 2HCl$$

The well-known example of the use of this synthetic pathway is Union Carbide's development of a process to manufacture methyl isocyanate by reacting methylamine with phosgene.[50] The reaction occurs as follows:

$$CH_3NH_2 + COCl_2 \longrightarrow CH_3NCO + 2HCl$$

The industrial use of this synthesis has proved to have terribly detrimental impacts on health and the environment, illustrated most graphically by the 1984 catastrophe in Bhopal, India, which involved the accidental release of the product, methyl isocyanate, and resulting in more than 15,000 fatalities.[22] Further processing of the isocyanate intermediate produced by this method involves transport and storage of large amounts. Moreover, both the major feedstock, phosgene, and the strong acid by-product, hydrochloric acid, are hazardous industrial chemicals.

4.5.1.4 Alternative Phosgene-Free Process to Methyl Isocyanate

In the mid-1980s DuPont developed an environmentally friendlier process useful for the industrial synthesis of methyl isocyanate based on the following reaction sequence.

$$CH_3NH_2 + CO \xrightarrow{\text{Catalyst}} CH_3NHCHO \longrightarrow CH_3NCO + H_2O$$

The DuPont process not only eliminates the use of phosgene as a starting material, but also avoids the production of large amounts of hydrochloric acid as an unwanted by-product. In this method, methylamine reacts with carbon monoxide to yield the corresponding aldehyde, which is then catalytically converted to isocyanate. This phosgene-free replacement synthesis also supports the trend in the chemical process industry to seek to reduce inventories on plant sites of hazardous synthetic reagents. Methyl isocyanate produced from this process is converted *in situ* to an agrochemical product.

Because the process is catalytic in the final step, and requires less than one kilogram of methyl isocyanate to be present (i.e., in inventory) to keep the operation running, it accomplishes the environmentally friendly goal of reduction in stock, storage, and handling of this hazardous chemical. This catalytic step can be done at the point of consumption, according to the source, by controlling the production of methyl isocyanate to meet the demand of the process stream.

By tailoring the design to achieve environmental benefits, several desirable goals have been achieved in this process advance. Not only is phosgene eliminated as a starting material, but unwanted production of hydrochloric acid as a by-product is avoided. Furthermore, the addition of a process-limiting, catalytically controlled step has drastically reduced the need for producing and storing large volumes of the highly toxic chemical intermediate, methyl isocyanate.

4.5.1.5 Synthesis of Urethanes, Isocyanates, and Ureas, Conventional Method

Carbamates (urethanes) and ureas are conventionally produced from amines via phosgene-based routes according to the following reactions:

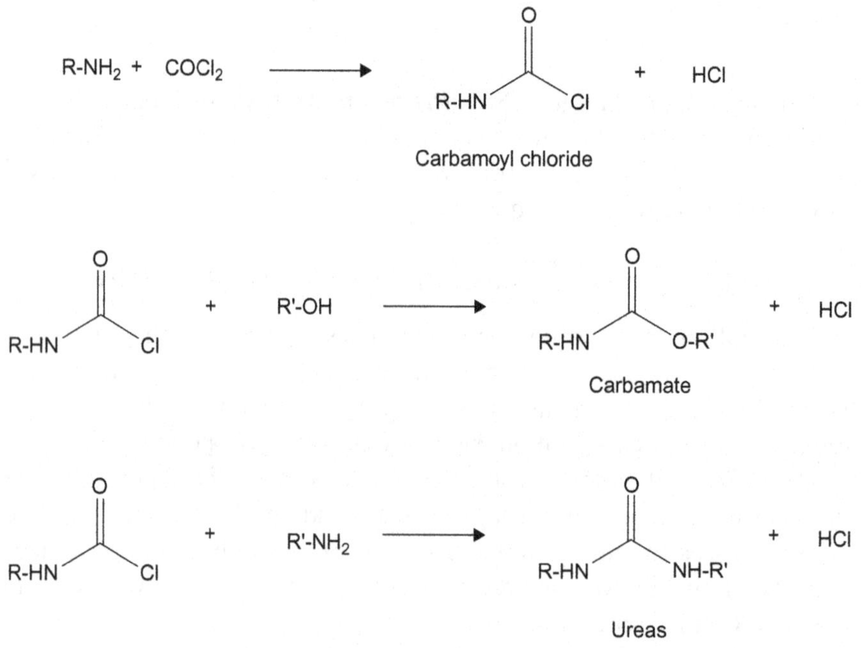

$$R\text{-}NH_2 + COCl_2 \longrightarrow \underset{\text{Carbamoyl chloride}}{R\text{-}HN-C(=O)-Cl} + HCl$$

$$R\text{-}HN-C(=O)-Cl + R'\text{-}OH \longrightarrow \underset{\text{Carbamate}}{R\text{-}HN-C(=O)-O\text{-}R'} + HCl$$

$$R\text{-}HN-C(=O)-Cl + R'\text{-}NH_2 \longrightarrow \underset{\text{Ureas}}{R\text{-}HN-C(=O)-NH\text{-}R'} + HCl$$

Reaction of alkylamine (RNH_2) with phosgene gives carbamoyl chloride, which further reacts with either alcohol or second alkylamines to produce corresponding carbamate or ureas, respectively. Dehydrohalogenation of carbamoyl chloride yields corresponding isocyanate as shown in the previous example.

4.5.1.6 Phosgene-free Synthesis of Urethanes, Isocyanates, and Ureas
4.5.1.6.1 EniChem's Dimethylcarbonate Technology
Just as described previously in the Daicel phosgene-free process for the production of isophorone diisocyanate (IPDI), this method uses dimethylcarbonate (DMC) as a phosgene substitute for the chemical starting block. EniChem Synthesis S.p.A (Italy) also uses this DMC technology in a wide variety of organic syntheses of pharmaceuticals, agrochemicals, dyes, polymers, and fine chemicals. Carbonylation reactions are commonly used to produce carbamates, isocyanates, and ureas as shown below:

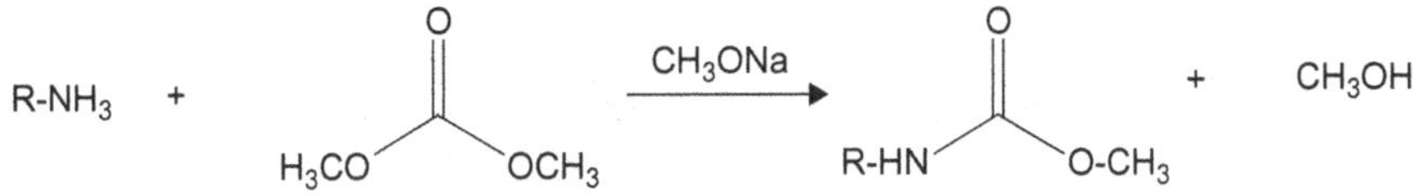

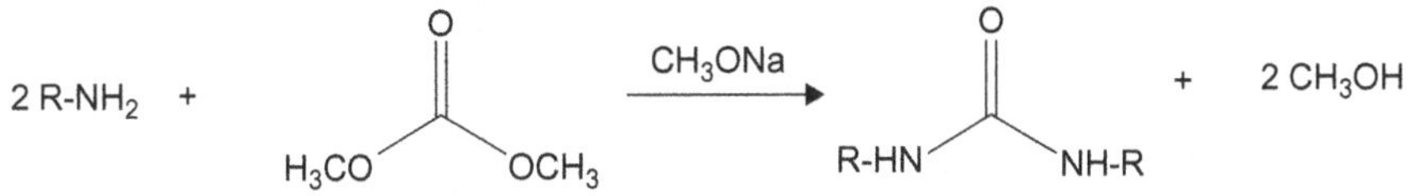

The formation of carbamates is also possible from hydrazines, imines, cyclic guanidines, and nitrogen heterocyclic compounds. For example, reaction of hydrazine with DMC gives methyl hydrazinoformate, which then further reacts with another molecule of hydrazine to yield carbonic dihydrazide (N,N'-diaminourea).

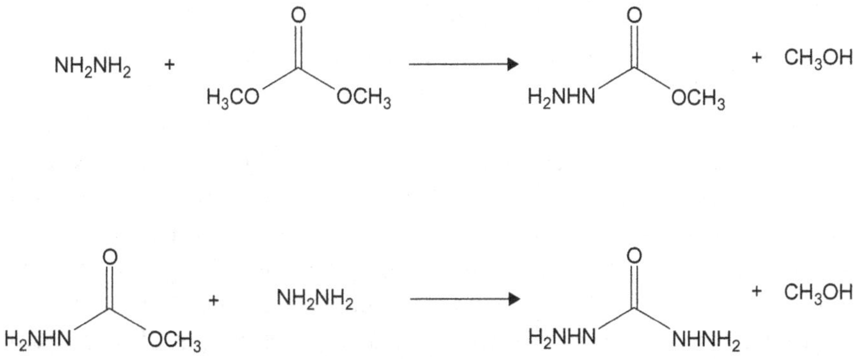

Such substitution also allows the chemical industry to eliminate the problems in removal and disposal of polluting byproducts, such as hydrochloric acid and chlorides, generated from the use of phosgene or chloroformates. DMC is produced by EniChem's patented process based on the oxidation of carbon monoxide with oxygen in methanol according to the following reaction:

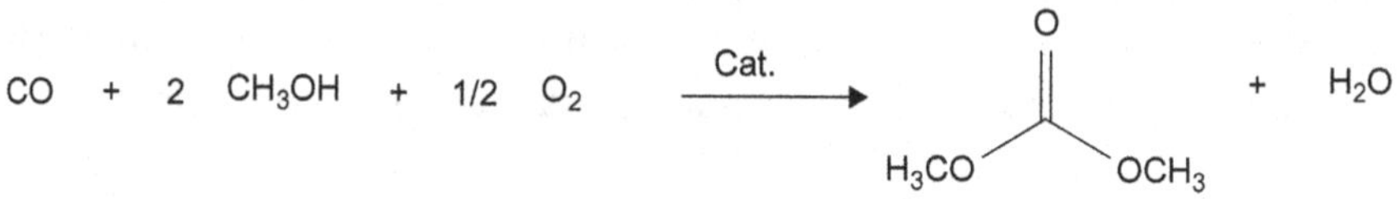

$$CO + 2\ CH_3OH + 1/2\ O_2 \xrightarrow{\text{Cat.}} \underset{H_3CO \quad\quad OCH_3}{\overset{O}{\underset{\|}{C}}} + H_2O$$

DMC made with this process is totally chlorine-free, whereas the conventional route involves the use of phosgene and methanol.

Isocyanates can be produced from carbamates (urethanes) through a pyrolysis reaction:

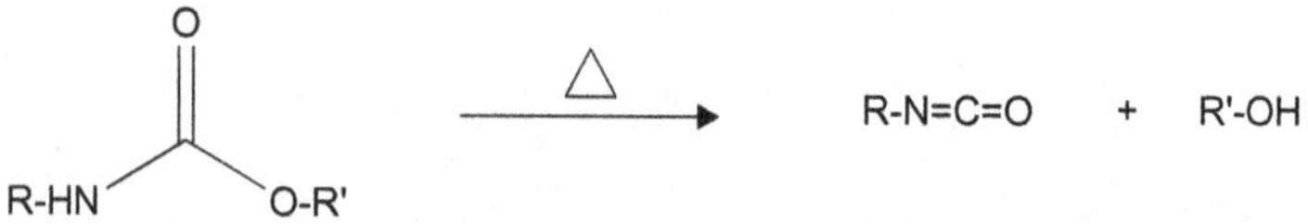

$$\underset{R\text{-}HN \quad\quad O\text{-}R'}{\overset{O}{\underset{\|}{C}}} \xrightarrow{\triangle} R\text{-}N{=}C{=}O + R'\text{-}OH$$

4.5.1.6.2 Monsanto's Phosgene-Free, Co-Based Process for Urethanes, Isocyanates, Ureas

The foundation of this synthetic pathway is replacement of the feedstock, phosgene, with carbon dioxide (CO_2). CO_2 can be activated by direct reaction with primary or secondary amines to produce the corresponding carbamate salts. These carbamate salts, derivatives of carbamic acid (NH_2COOH), release the carbamate anion ($RNHCOO^-$), which can be activated in a controlled way to generate isocyanates or urethanes. It has been shown in a number of documented reactions that electrophilic attacks on the carbamate anion generally occur at the N-position. Thus, reactions of carbamate ions with an alkyl halide produce secondary or tertiary amines and reactions with acylating agents produce amides. With the use of controlling conditions, Monsanto chemists have developed a synthetic method in which the focus of electrophilic attack is redirected from the nitrogen (N) to the oxygen (O) nucleophilic center of the anion. This specific activation of the carbamate oxygen site is accomplished by utilization of a carefully selected tertiary amine base, which is unreactive toward CO_2 under the mild conditions employed in the reaction. This CO_2-based technology was found to be a good source for direct production of urethanes, according to the following reaction sequence:

$$RNH_2 + CO_2 + R_3'N \longrightarrow R_3'NH^+RNHCOO^-$$
$$(R_3'NH)^+(RNHCOO)^- + EX \longrightarrow RNHCOOE + R_3'NH^+X^-$$

where EX = an electrophile, for example, an alkyl halide

Subsequently, the Monsanto researchers developed a variation of this "activated carbon dioxide chemistry" process whereby activated carbamate anions derived from primary amines could be reacted rapidly with electrophilic dehydrating agents, such as acid halides, to produce the corresponding isocyanates in excellent yields, according to the following reaction:

$$RNHCOO^-R_3NH^+ + RCOCl + R_3N \longrightarrow RNCO + R_3NH^+Cl^- + RCOOH$$

This has proven to be a very versatile reaction, which can start with a variety of electrophilic "dehydrating" agents and organic bases. Furthermore, the mild reaction conditions allow for the use of amine precursors, which can possess a number of different functional groups. Thus the Monsanto activated carbon dioxide process not only eliminates the use of phosgene as a starting material, but also provides additional benefits, including milder reaction conditions, urethane and isocyanate products in quantitative yields, reductions in problem impurities and by-products, and broader possibilities in choice of amine feedstock.

4.5.1.7 Phosgene-Free Production of High Molecular Weight Polycarbonates

4.5.1.7.1 Conventional Route for Industrial Scale Production of Bisphenol-Based Copolymer Used in Manufacture of Polycarbonate Resins

A characteristic example of the phosgene-based process for manufacture of polycarbonates is the synthetic method used by Dow Chemical Co. A typical variation of the process uses bisphenol A, diphenyl carbonate, phosgene, cuprous chloride, and an oxygen-containing gas. Thus, when a phenol is reacted with phosgene, a diaryl carbonate is formed, which is then reacted with a bisphenol monomer to obtain a bisphenol polycarbonate and phenol, the latter of which after separation from the polycarbonate is reacted with more phosgene, and the diaryl carbonate thus produced is cycled back into the process for reaction with the bisphenol monomer.

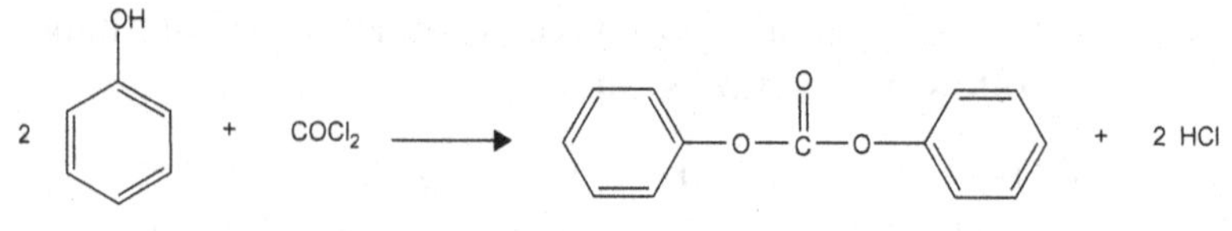

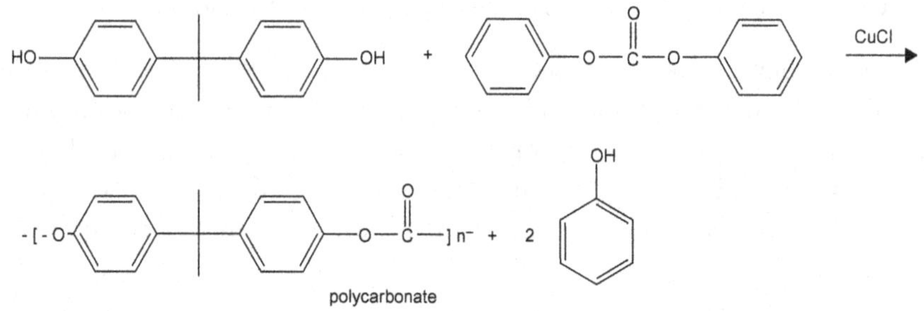

polycarbonate

4.5.1.7.2 Daicel Phosgene-Free Process to Produce Copolymer Feedstock for Polycarbonates

In a Japanese patent assigned to Daicel Chemical Industries, Ltd,[51] a process is described for the non-phosgene synthesis of a bisphenol-based copolymer to be further processed for the manufacture of high molecular weight polycarbonates. In this process, bisphenol A diacetate was reacted with dimethyl carbonate in the presence of potassium borohydride as a catalyst and autoclaved at 160 °C to produce the oligomer in high yield. This oligomer, in turn, was heated in the presence of the same catalyst *in vacuo* at a temperature of 240–280 °C yielding a polycarbonate with an average molecular weight of 35,000.

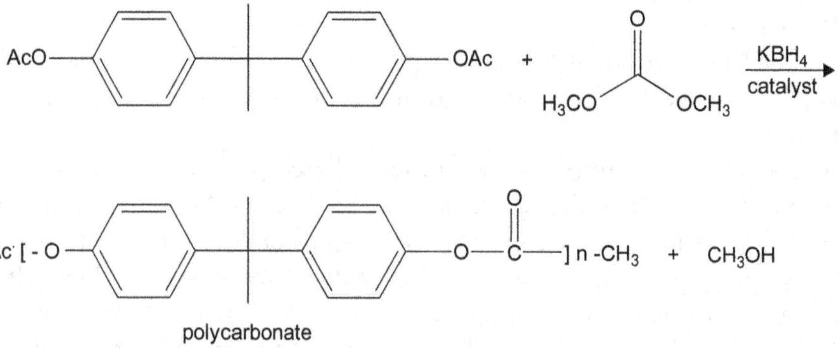

polycarbonate

The foregoing examples of phosgene-free methods for synthesis of carbamates, isocyanates, ureas, polycarbonates, dyes, polymers, and other pharmaceutical or agrochemical intermediates and products all implicit in phosgene production, handling, storage, and transport. But perhaps even more importantly, these same green chemistry methods also eliminate the possibility that those facilities will be targeted by terrorists or be the source of an uncontrolled release of what are, historically, some of the most toxic gases in industry. It should be noted that there are significant data showing that BPA is an endocrine disruptor, and its inclusion in this section should not be interpreted as an endorsement of its use.

4.6 SULFUR DIOXIDE (SO₂)

Background: Sulfur dioxide is a colorless gas with a sharp pungent odor. It may be shipped and stored as a compressed liquefied gas. Sulfur dioxide is used in the manufacture of sulfuric acid, sulfur trioxide, and sulfites. It is also used in solvent extraction and as a refrigerant. Acute exposure irritates the eyes and air passages. High concentration exposures to the skin and eyes can cause severe burns and blindness, and breathing high concentrations can lead to permanent lung damage and death.

The gas form is heavier than air and can quickly reach concentrations that are harmful when inhaled.

Sulfur dioxide is among the top 10 chemicals involved in industrial accidents as reported by the EPA's Risk Management Planning program.

4.6.1.1.1 Anhydrous Sulfur Dioxide in Pollution Control
Wisconsin Power Pulliam Plant, in Green Bay, WI, switched from anhydrous sulfur dioxide, used to capture particulates in pollution control equipment, to a safer solid form of the chemical. The change eliminated potential off-site injury to any of 180,000 people.

4.6.1.2 Sulfur Dioxide in Food Products
4.6.1.2.1 Sodium Bisulfite Replaces Sulfur Dioxide in Corn-Milling
Cargill, Inc. plants in Memphis, TN, and Eddyville, IA, manufacture products, such as corn oil, corn syrup, and animal feed from corn. These plants formerly used anhydrous sulfur dioxide to soak and soften corn kernels in the corn-milling process. Both switched to the less hazardous—but still effective—sodium bisulfite as a replacement. This practice eliminated off-site vulnerability to 19,000 people in Eddyville and 370,000 people in Memphis. The Environmental Manager of the Memphis plant was quoted as saying, "Switching to the safer sodium bisulfite is a good best practice for the industry."[29]

4.6.1.2.2 Sodium Bisulfite Replaces Sulfur Dioxide in Soy Processing
Solae Company, dba DuPont Soy Polymers (formerly Protein Technologies International), in Louisville, KY, extracts protein from soybean flakes for use in products, such as soy flours, concentrates, and isolates. The facility formerly used anhydrous sulfur dioxide to bleach products, stabilize drying, and lower pH. To improve safety, Solae switched to sodium bisulfite, a less acutely hazardous chemical. The change improved the safety of more than 37,000 residents and others who work in Louisville.

4.7 SULFURIC ACID (H$_2$SO$_4$)

Background: Sulfuric acid is a highly corrosive, water-soluble, strong acid; a 0.50 M solution of sulfuric acid has a pH close to zero. Though not considered toxic per se, the main occupational risk is skin contact, leading to burns and inhalation of aerosols, which, at high concentrations, leads to immediate and severe irritation of the eyes, respiratory tract, and mucous membranes. The corrosive properties of sulfuric acid are accentuated by its highly exothermic reaction with water, and burns from sulfuric acid are potentially more serious than those from comparably strong acids (e.g., hydrochloric acid), because there is additional tissue damage due to the heat liberated by the reaction with water, producing secondary thermal damage. It is also a principal constituent of acid rain.

Sulfuric acid has many applications, and is produced in greater amounts than any other chemical, besides water. World production in 2001 was 165 million metric tons, with an approximate value of $8 billion. Principal uses include ore processing, fertilizer manufacturing, detergent manufacturing, oil refining, wastewater processing, and chemical synthesis. It is used in large quantities in iron and steel making, principally to remove oxidation and rust.

4.7.1.1 Sulfuric Acid in Cleaning Products
4.7.1.1.1 Just-in-Time Manufacturing Greatly Reduces
Volumes on Hand
The Proctor and Gamble Company Alexandria Plant, Pineville, La., makes surfactants for detergents and dry laundry products. Previously, the company purchased oleum (fuming sulfuric acid)

from another company. As part of a major plant upgrade, the facility installed a sulfur-burning unit that makes sulfur trioxide on demand for immediate use. This "just-in-time" production eliminated the need to transport and store large quantities of oleum. The new production method eliminated the danger of a chemical release to some 2200 residents in the community, as well as to schools, churches, and a Wal-Mart nearby.[29]

4.7.1.2 Sulfuric Acid in Organic Synthesis and Resulting Inorganic Salt Waste

With increasing economic and environmental pressure for the chemical industry to minimize or, preferably, to eliminate waste generation during the product manufacturing process, much effort has been devoted to identifying the environmentally friendly alternatives that could lead to "zero waste" and "zero emission" of pollutants. A large volume of waste generated from manufacturing processes is inorganic salts that result from acid-base neutralizations. The most common salts are sodium chloride, sodium sulfate, and ammonium sulfate. Such inorganic salts require either disposal or recycling/regeneration treatment, which presents a great burden in terms of cost, as well as liability problems for chemical manufacturers.

The conventional acetone cyanohydrin (ACN) process for methyl methacrylate synthesis described above for hydrogen cyanide is a typical illustration of waste-salt generation resulting from use of sulfuric acid in the manufacturing process. Not only does this process involve use of hydrogen cyanide and sulfuric acid, it also produces a very large quantity of ammonium bisulfate in the waste acid stream that requires further treatment and disposal. By applying novel catalytic technologies in the synthesis of methyl methacrylate, the desired product can now be commercially made without generating waste-salts.

The following four examples (caprolactam, resorcinol, hydroquinone, and gasoline alkylation technology) also involve the similar application of clean catalytic technologies in organic synthesis to replace the use of sulfuric acid in the traditional process, thereby reducing the hazards implicit with handling, storing, and transporting large quantities of sulfuric acid.

1. *Conventional method for manufacture of caprolactam*

Caprolactam is a key raw material for nylon-6 manufacture. The conventional route using sulfuric acid to produce caprolactam is shown below:

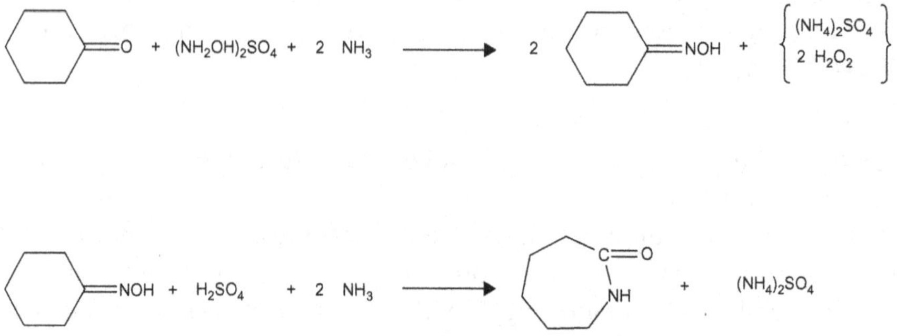

Reaction of cyclohexanone with hydroxylamine-which is produced as a sulfate salt by the air oxidation of ammonia to nitric acid, followed by catalytic hydrogenation in the presence of sulfuric acid and ammonia yields an intermediate, cyclohexanone oxime, which undergoes Beckmann rearrangement to caprolactam in the presence of stoichiometric amounts of sulfuric acid. Large quantities of salt are produced in both reactions, making it a very wasteful process: Approximately 4.5 kg of ammonium sulfate salt is generated per kilogram of caprolactam that is produced.

2. *Sulfuric acid-free manufacture of caprolactam*

An alternative route to cyclohexanone oxime developed in Italy by Enichem is shown in the following reaction. Cyclohexanone oxime is produced by the ammoxidation of cyclohexanone with ammonia and aqueous hydrogen peroxide in the presence of a solid, recyclable catalyst, titanium silicalite (TS-1). This reaction step eliminates approximately one-third of total salt formation. However, the oxime is still converted to caprolactam through the conventional route (Beckmann rearrangement), catalyzed by stoichiometric amounts of sulfuric acid, and produces ammonium sulfate salt. Therefore, this alternative process still leaves something to be desired.

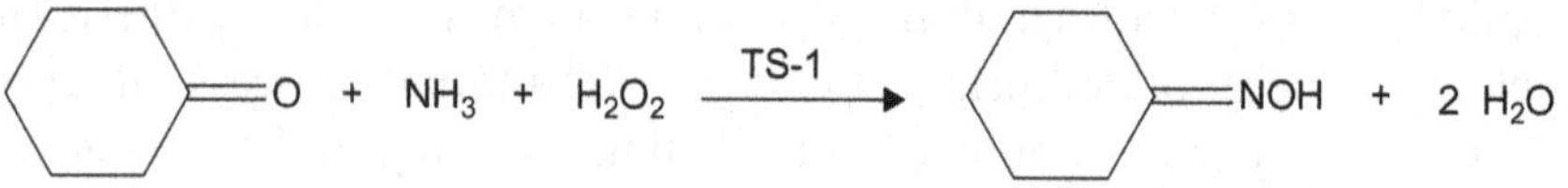

However, another approach to find the suitable catalyst for the conversion of cyclohexanone oxime to caprolactam in order to completely eliminate the salt formation has been reported by Sumitomo, of Japan. They reported the use of a solid high-silica zeolite catalyst (ZSM-5) for the gas-phase rearrangement of cyclohexanone oxime at 350 °C. Caprolactam is produced with 95% selectivity at 100% oxime conversion.

Union Carbide also reported the use of a proprietary catalyst, SAPO-11, for the same reaction. Successful application of these catalytic technologies can reduce unwanted salt generation and replace sulfuric acid in these processes.

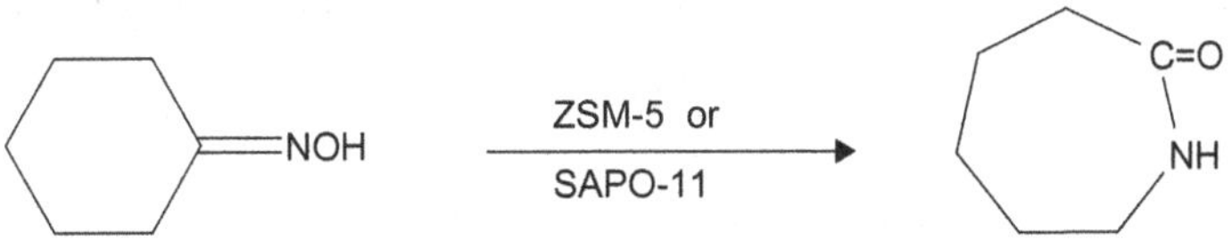

3. *Conventional method for manufacture of resorcinol:*

Resorcinol is commonly produced by the benzene disulfonation process as shown below. This process produces two equivalents of sodium sulfite (Na_2SO_3) salt waste per equivalent of product.

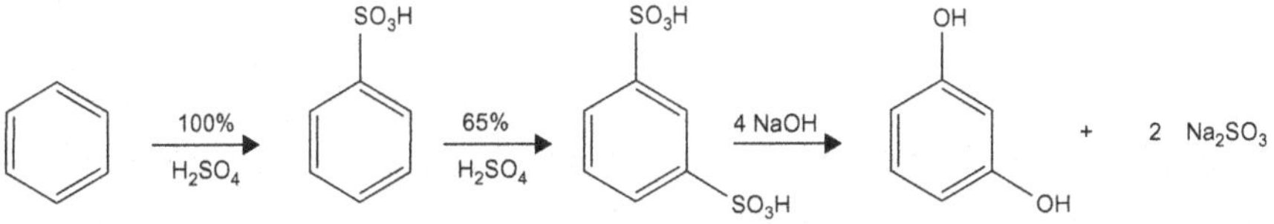

4. *Resorcinol manufacture without sulfuric acid:*

Sumitomo of Japan has developed a commercially alternative process that avoids the use of sulfuric acid and the concomitant undesired salt formation (see reaction sequence below). This alternative route is analogous to the cumene process for phenol production. Diisopropylbenzene (m-DIPB) can be manufactured by the catalytic dialkylation of benzene with two equivalents of propylene. The resulting m-DIPB is then catalytically hydroperoxidized to the corresponding dihydroperoxide (DHP). Upon acidification the dihydroperoxide (DHP) is cleaved cleanly to give resorcinol and acetone. It is worth noting that the Sumitomo process not only provides a cleaner alternative route for the production of resorcinol, but also generates acetone as a useful byproduct, whereas the conventional route generates unwanted waste salt.

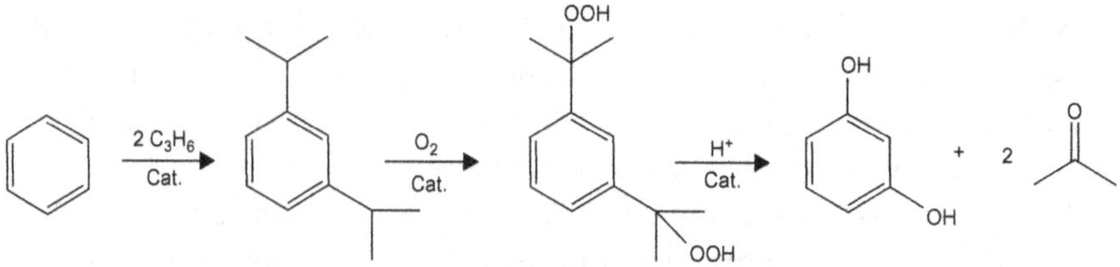

4.7.1.2.1 Conventional Method for Manufacture of Hydroquinone

The zeolite-based titanium silicalite-1 catalyst system is also applied in the production of hydroquinone and catechol. The following reaction scheme shows the conventional route for the manufacture of hydroquinone.

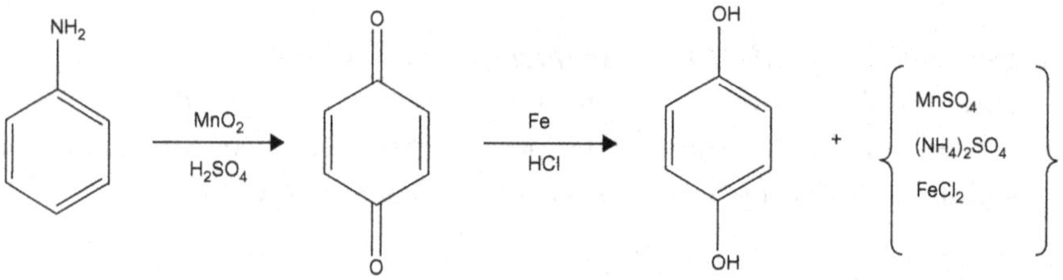

The widely used conventional route involves aniline oxidation and quinone reduction reactions. Aniline is oxidized to quinone by using excess amounts of manganese dioxide (MnO_2) in sulfuric acid. The resulting quinone is steam-stripped from the oxidation solution, and then reduced to hydroquinone by acidic iron solution (Fe/HCl). The total yield of hydroquinone is 90% (based on aniline), but substantial amount of manganese sulfate ($MnSO_4$) and ammonium sulfate ((NH_4)$_2SO_4$) salts are generated from the aniline oxidation step, as well as iron(II) chloride ($FeCl_2$) salt from the quinone reduction step.

4.7.1.2.2 Hydroquinone Manufacture without Sulfuric Acid

An alternative route to hydroquinone has been developed independently by Snamprogetti, Enichem, and National Chemical Lab Phenol can be directly hydroxylated to hydroxyquinone and catechol with hydrogen peroxide over titanium silicalite zeolites catalyst, without reliance on sulfuric acid:

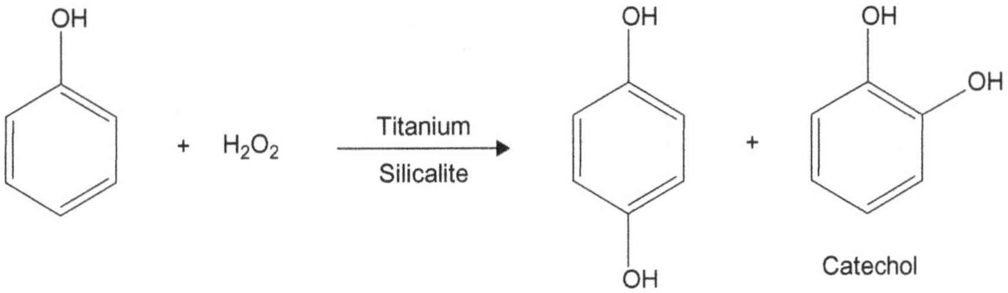

Catechol

Another catalytic route has also been considered:

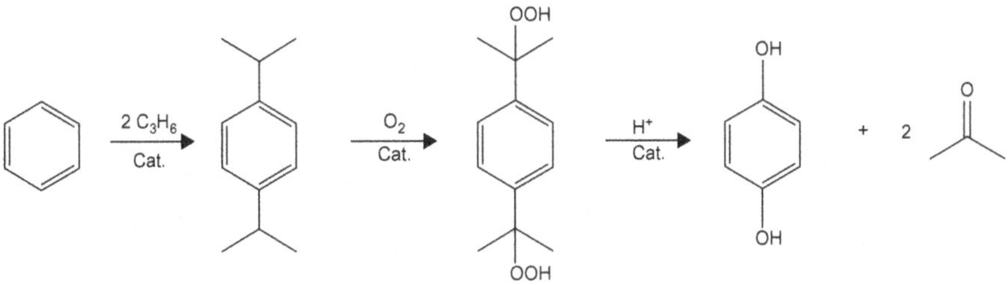

This alternative route is similar to the process for resorcinol production, the only difference being the use of the respective diisopropylbenzene (DIPB) isomers. Hydroquinone or resorcinol is produced from p-DBPB or m-DIPB, respectively, according to the dihydroperoxide reaction. In other words, hydroquinone and resorcinol can be produced alternatively in the same manufacturing plant when the reaction processes are designed for alternatively producing the m- and p-DIPB isomers. In addition, diisopropylbenzenes (DIPBs) are by-products from the process of cumene production in phenol manufacture.

These examples demonstrate a good economic practice for the chemical industry. By applying these catalyst technologies on manufacturing processes, the chemical industry can produce important chemical building blocks, such as cumene, phenol, hydroquinone, and resorcinol in the same plant from a common feedstock (see following scheme). It not only offers advantages of eliminating environmentally unfriendly chemicals, such as sulfuric acid and inorganic waste salts, but also provides tremendous economic efficiency, and is an excellent example of green chemistry or "benign-by-design" practices.

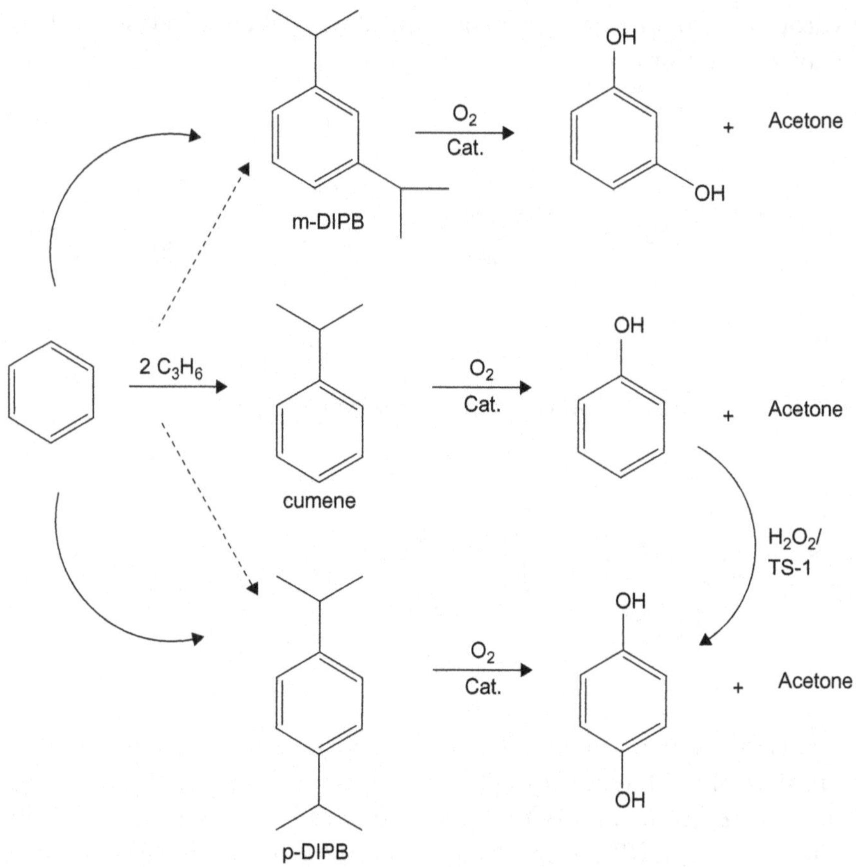

4.7.1.3 Gasoline Alkylation Technology

A detailed discussion of the conventional and alternative liquid acid catalysts used in the production of gasoline alkylate is described under the section on alternatives to hydrogen fluoride.

The traditional sulfuric acid-based process is comparatively safer than the analogous HF-based process, but it too generates large quantities of spent acid that need to be continuously removed from the process, regenerated, and replaced.

It is also worth noting that the sulfuric acid used in alkylation plants represent the largest quantity of catalyst used in any chemical and refining process, so it requires enormous on-site inventories, and also presents handling and transportation problems.

4.8 AMMONIA

Background: Ammonia (anhydrous) is a corrosive colorless gas with a strong odor. It is used in making fertilizer, plastics, dyes, textiles, detergents, and pesticides. Acute ammonia exposure can irritate the skin and burn the eyes, causing temporary or permanent blindness. It can also cause headaches, nausea and vomiting. Prolonged exposure to concentrations greater than 300 ppm can cause fluid in the respiratory system (pulmonary or laryngeal edema), which may lead to death. Chronic exposure damages the lungs; repeated exposure can lead to bronchitis with coughing or shortness of breath. It is somewhat flammable at concentrations of 15–28% by volume in air. Fortunately, ammonia has a low odor threshold (5 ppm), and causes eye irritation at 20 ppm, alerting most people to its presence and the need to move out of the area of contamination.

Ammonium nitrate requires little skill in chemistry to convert to a useful explosive, although the putative terrorist would also need to construct a detonation system. At about US $750/tonne in bulk, it is relatively inexpensive.

Anhydrous ammonia was the number one most frequently involved chemical in accidents documented for the RMP*Info database[32] (see Table 2.1).

4.8.1.1.1 Ammonia in Glass Production
PPG Industries, in Fresno, CA, manufactures flat glass used in windows and architectural applications. In 2000, the facility went from air natural gas combustion to oxygen natural gas combustion, called "oxyfuel." Using this different firing method eliminated the need for anhydrous ammonia in pollution control. The change was part of a larger $40 million upgrade that reduced nitrous oxide emissions to meet air quality requirements. In addition, the company realized improved manufacturing efficiency and product quality, while eliminating the danger of anhydrous ammonia formerly posed to some 14,300 nearby residents.

AFG Industries, in Victorville, CA, a manufacturer of flat glass, formerly used an ammonia injection system to control nitrous oxide emissions. This system required storing anhydrous ammonia. To further reduce air emissions from glass furnaces, the company adopted a natural gas process (Pilkington 3R technology). The change eliminated a vulnerability zone of 82,000 people.

4.8.1.1.2 Ammonia in Chemical Manufacturing

Calgon Carbon Corporation, in Neville Island Plant, Pittsburgh, PA, produces activated carbon for use in respirators and other products. The company previously treated the carbon with aqueous ammonia that was produced on-site from anhydrous ammonia. The company retained the same carbon treating process, but now starts with the aqueous ammonia. Savings on safety and security compliance offset slightly increased shipping costs. The change eliminated a vulnerability zone that formerly encompassed 120,000 people.

4.8.1.1.3 For More on Ammonia...

For more examples of synthetic approaches that have reduced the use of ammonia, see also the case studies describing: catalytic dehydrogenation of diethanolamine, HCN-free synthesis of the amino acids, and HCN-free synthesis of the specialty chemical, sarcosinate, all listed under the section on hydrogen cyanide, and the case study describing synthesis of the drug aprepitant.

4.9 METHANE (CH$_4$) CONVERSION

Background: Methane is a highly flammable, colorless, odorless gas at ambient temperatures. It is the principal constituent of natural gas (85%), and is also a major greenhouse gas found in the atmosphere. It is released to the environment as natural emissions from microbes, animal waste, volcanoes, the rumen of domestic animals (especially cattle), and during the growing of rice. Methane is stored as methane hydrates in immense amounts, both in Arctic regions and in marine sediments, and the worldwide amounts of methane hydrates are conservatively estimated to total twice the amount of carbon found in all known fossil fuels on Earth, making it an important potential source of fuel energy.

Although world reserves of methane-rich natural gas are abundant, transporting the volatile gas from its remote sites of origin to sites of chemical manufacture is expensive and dangerous. The object of converting methane to a more easily transportable liquid has long been a goal of the chemical industry.

Conversion of methane to methanol, for example, would allow methane to equal petroleum as a feedstock for multiple chemical uses. Though methane is not toxic, it poses an immediate health hazard in that it can cause burns if it ignites, and can form explosive mixtures with air. Methane is the most abundant of all alkane compounds, yet the hazards of its handling and distribution prevent exploitation of known methane reserves.

In addition to its potential use as a feedstock for multiple chemical syntheses, conversion of methane to methanol is of great benefit as a direct route to produce oxygenated fuels. Methanol may be used in direct combination with gasoline or as a raw material for methyl tertiary butyl ether for mixing with gasoline. The catalytic conversion of methane to methanol or acetic acid has enormous potential to:

1. Upgrade the economic potential of vast unexploited reserves of methane-rich natural gas.
2. Allow transport of methane safely by converting it from a volatile explosive gas to a much less volatile and easily transportable liquid.

What follows are descriptions of a conventional method for converting methane to methanol, followed by several alternative approaches.

4.9.1.1 Conventional Route for Methane Conversion
The classical route of methane to methanol conversion involves oxidation of methane using various conditions. However, until recently, no satisfactory method had been devised to control the oxidation sufficiently to stop the reaction at methanol. Instead, oxidation tended to proceed beyond methanol to produce unwanted carbon dioxide. The conventional steam reforming process, which produces the intermediate syngas (a mixture of hydrogen gas and carbon monoxide, from which methanol can be synthesized), comprises 60–70% of the capital costs of methane to methanol conversion. Thus, a low temperature catalytic process that avoids syngas production would be highly advantageous from an economic perspective.

Conventional methods of methane to methanol conversion have produced poor yields. A summary of conventional routes of methane to methanol conversion is presented in Figure 4.1.

4.9.1.2 Alternative Routes of Methane Conversion
1. *Catalytic process for methane-methanol conversion (joint research with Petro-Canada + Mitsubishi)*
   Catalytica has a patented process for methane-methanol conversion using low temperature homogeneous catalysis. Under relatively

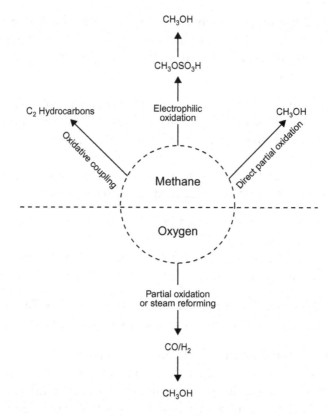

Figure 4.1 Conventional Methane—Methanol Conversion Routes[8].

moderate temperatures of 180 °C, metal catalysts and sulfuric acid react with methane to form a methyl ester, which is then hydrolyzed to form methanol. While the initial work was performed using a mercury catalyst, Catalytica has reported comparable results using non-mercury catalysts as well, including palladium, platinum, and gold.

Although this process offers the advantage of driving a low temperature, carefully controlled oxidation of methane, thereby increasing the yield of methanol, it also utilizes sulfuric acid to produce the intermediate methyl bisulfate. The need for acid resistant containers to perform these reactions may raise costs of the process. And although the sulfuric acid is recovered and recycled into the process, the environmental benefits of this methane conversion are somewhat offset by the need to ship and store hazardous sulfuric acid. The trade-off between safer methane transport versus increased sulfuric acid transport and storage needs to be considered from the perspective of accidental releases.

2. *Methane-Methanol Conversion. University of Minnesota process to catalytically oxidize methane to syngas without steam reforming:*
Researchers at the University of Minnesota developed a novel process that directly converts methane to syngas (a mixture of hydrogen gas and carbon monoxide, from which methanol can be synthesized), bypassing the usual steam reforming step. In this process, a rhodium-based catalyst is utilized to produce syngas at ambient temperatures and very short contact times:

$$CH_4 \xrightarrow[O_2]{\text{Rhodium}} CO/H_2 \xrightarrow{\triangle} CH_3OH$$

However, the high temperature required for methanol synthesis may still make it difficult to achieve high yields.

3. *Methane-acetic acid conversion via direct catalytic conversion:*
Researchers at the Pennsylvania State University have reported direct catalytic conversion of methane into acetic acid under relatively mild conditions. Conventional synthesis of acetic acid from methanol currently involves a three stage process, including:

1. High temperature steam reforming of methane to a 3:1 mixture of syngas (3 parts CO to one part H_2);
2. High temperature conversion of a 2:1 mixture of syngas to methanol; and
3. Carbonylation of methanol to acetic acid (primarily using the Monsanto Process).

The Monsanto carbonylation process is catalytic using rhodium and iodine:

$$CH_3OH + CO \xrightarrow{Rh\,/\,I_2} CH_3COOH$$

In the University of Pennsylvania process, the reaction proceeds via catalytic oxidation of methane using oxygen and a rhodium trichloride catalyst:

$$CH_4 + CO + 1/2\,O_2 \xrightarrow[\substack{H_2O \\ 100°C}]{RhCL_3} CH_3COOH$$

The only byproducts observed were methanol and formic acid. While this reaction is a novel one (it is the first time oxygen has been used as the methane oxidant), it remains unclear whether it has useful commercial application. The slow rate of the reaction is a handicap, although use of other catalysts may ultimately speed up reaction rates. This research was initiated as part of the search for efficient methane-methanol conversion routes, and may yet have some applicability to this purpose, although the reaction probably proceeds without methanol formation as an intermediate step (as evidenced by conversion of tagged methanol to formic acid, and not acetic acid under the conditions of the experiment).

This example offers potential for acetic acid conversion from methane. The current market for acetic acid is valued at $4.5 billion.[52] Exploitation of methane reserves can more safely occur following conversion by the easy transport of acetic acid, rather than gaseous methane.

Because methane is such an explosive—yet valuable—fuel and feedstock, any green chemistry methods by which it can be converted to methanol will greatly increase its viability as a useful input for industry, particularly by reducing the risks associated with handling and storing it. Conveniently collateral to this, there will be a reduced risk that methane storage and transport facilities could present a tempting target for terrorists to attack.

4.10 SOLVENTS AND SOLVENT REDUCTION

4.10.1.1 Greener Solvents and Reclamation in the Flexographic Printing Industry

Flexographic printing is used in a wide array of printing, but uses millions of gallons of solvent. Arkon Consultants and NuPro Technologies, Inc. have developed a safer chemical processing system that eliminates hazardous solvents, reduces explosion potential and emissions, and increases worker safety in the flexographic printing industry.

Flexographic printing is used on everything from food wrappers to secondary containers, such as cereal boxes to shipping cartons. The photopolymerizable material on a flexographic printing plate cross-links when exposed to light and captures an image. After exposure, flexographic printing plates are immersed in a solvent to remove the unpolymerized material. The developing, or washout, solvent is typically a mixture of chloro, saturated cyclic, or acyclic hydrocarbons. Xylene is the most common solvent. Most traditional washout solvents are hazardous air pollutants (HAPs) subject to stringent reporting requirements; they also raise worker safety issues and create problems with recycling and disposal. North America alone uses two million gallons of washout solvents each year with a market value of $20 million. Many small printing plants also use these solvents.

NuPro/Arkon have developed a safer chemical processing system, including several new classes of washout solvents with methyl esters, terpene derivatives, and highly substituted cyclic hydrocarbons. The advantages include higher flash points and lower toxicity, which reduce explosion potential, worker exposure, and regulatory reporting. The methyl esters and terpene derivatives are biodegradable, and can be manufactured from renewable sources. All of their solvents are designed to be recycled in their Cold Reclaim System™. In contrast to traditional vacuum distillation, this combination of filtration and centrifugation lowers exposures, decreases maintenance, and reduces waste. The waste is a solid, nonhazardous, polymeric material.

In the U.S. market, NuPro/Arkon are currently selling washout solvents that are terpene ether- and ester-based or made with low-hazard cyclics, and are marketing their methyl ester-based solvent in China and Japan. Their first filtration-based Cold Recovery System™ is currently in use in Menesha, WI, and is being marketed to larger U.S. users. Their centrifugation reclamation system for smaller users is in the final stages of development.

Use of these solvents and systems benefits both human health and the environment by lowering exposure to hazardous materials, reducing explosion potential, reducing emissions, and, in the case of the terpene and methyl ester-based solvents, using renewable resources. These solvents and the reclamation equipment represent major innovations in the safety of handling, exposure, and recovery. The reduced explosion potential, reduced emissions, decreased worker exposure, and reduced transport and maintenance costs translate into decreased cost and improved safety in all aspects of flexographic printing processes.

By replacing the need for xylene, chloro, saturated cyclic, or acyclic hydrocarbon solvents and offering up new, safer alternatives for this 2 million gallon per year industry in the U.S. alone, this technology is also reducing the likelihood that these facilities could constitute a viable target for attack.

4.11 BENZENE (C_6H_6)

Background: Benzene is a colorless and flammable liquid with a sweet smell and a relatively high melting point. Benzene is a natural constituent of crude oil, but it is usually synthesized from other compounds present in petroleum. It is a known human carcinogen, is no longer used as an additive in gasoline, and its functions as a solvent have largely been replaced in the laboratory by toluene. But it continues to be an important industrial solvent and precursor in the production of other chemicals. Its most widely-produced derivatives include:

1. Styrene, which is used to make polymers and plastics;
2. Phenol, for resins and adhesives; and
3. Cyclohexane, which is used in the manufacture of nylon and nylon resins.

Smaller amounts of benzene are used to make some types of rubbers, lubricants, dyes, detergents, drugs, explosives, and pesticides.

Figure 4.2 (a) Smoke billows out after a series of blasts hit a chemical factory in Jilin City, Northeast China's Jilin Province Sunday, November 13, 2005. (b) Note the proximity to urban residential neighborhoods and the wide spread of contamination to surrounding areas.[55]

Breathing high levels of benzene can result in death, while low levels can cause drowsiness, dizziness, rapid heart rate, headaches, tremors, confusion, and unconsciousness. Eating or drinking foods containing high levels of benzene can cause vomiting, irritation of the stomach, dizziness, sleepiness, convulsions, rapid heart rate, and death.[53]

Water and soil contamination are important pathways of concern for transmission of benzene contact. In the U.S. alone there are approximately 100,000 different sites which have benzene soil or groundwater contamination (Figure 4.2).[54]

In November 2005, a petrochemical plant in Jilin City, China suffered a series of explosions over the course of an hour when a nitration unit for aniline equipment jammed up, killing at least five, injuring about 70, and causing the evacuation of tens of thousands of residents. The blasts also created an 80 km long toxic slick of benzene and nitrobenzene in the Songhua River, which took weeks to clear. The water supply to the city of Harbin in China, with a population of almost nine million people, was cut off because of the severity of the benzene exposure.[56]

4.11.1.1 Benzene in Production of Styrene

A number of important products for which benzene is currently used may be manufactured using inherently safer green chemistry. An example of this is styrene. Styrene is widely used as a monomer for the production of polystyrene and ABS (Acrylonitrile-Butadiene-Styrene) resin. Polystyrene is commonly used in packaging and disposable service-ware materials, such as food/fast food packaging, furniture, appliances, electrical and electronic equipment, housewares, and various related consumer products. The main uses of ABS resin in industrial application include interior and exterior automobile parts, drain/waste/vent pipes, appliance parts, luggage, computer consoles, and other household materials. Styrene is also used as a feedstock to produce some other commercially significant styrene-derivative chemicals, such as divinylbenzene, vinyltoluene, and α-methylstyrene.

4.11.1.2 Conventional Benzene-Based Routes to Produce Styrene

Currently, there are two commercially available routes for the production of styrene monomer. Both routes involve the use of ethylbenzene as starting material, and ethylbenzene was first produced by the classical Friedel-Crafts alkylation of benzene with ethylene in the presence of an aluminum chloride catalyst complex, as shown in the following reaction:

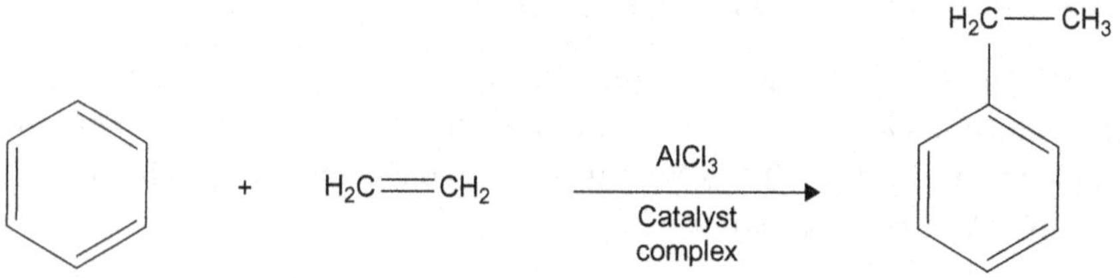

Some modifications of this method have been developed by companies including Dow Chemical, BASF, Shell Chemical, Monsanto, and Union Carbide. Most of their efforts were focused on the development of various complex catalyst systems based on liquid-phase (aluminum chloride-based catalysts) and vapor-phase (solid acid-based catalysts, such as phosphoric acid catalysts) processes.

During the 1970s, a new ethylbenzene process (the Mobil/Badger process) was developed by Mobil Oil Corporation. This process was based on a zeolite catalyst, ZSM-5, that provided not only high catalytic activity (with relatively good resistance to coke formation), but also quantitative product yields. Since then, this technology was widely applied in the production of ethylbenzene. The major advantages of using this new process include the following:

1. The catalyst system (silica-alumina-based catalyst) is non-corrosive and truly heterogeneous;
2. The costly facilities associated with traditional liquid-phase process are eliminated; and

3. Aqueous wastes (resulting from aluminum chloride catalyst) and waste-treatment facilities are eliminated.

However, it still can not avoid the use of benzene as the feedstock for all of the ethylbenzene processes. Styrene monomer is conventionally produced from ethylbenzene by the following processes:

4.11.1.2.1 Conventional Dehydrogenation of Ethylbenzene

Styrene is produced by direct dehydrogenation of ethylbenzene in the presence of the dehydrogenation catalyst and steam (heat) as shown in the following reaction:

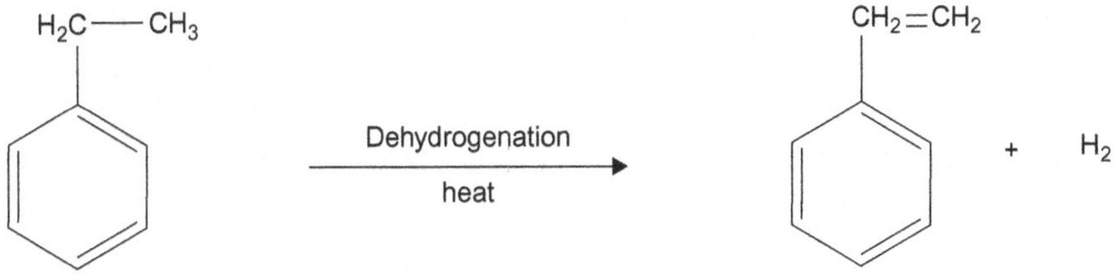

The dehydrogenation reaction is an endothermic reaction, with the heat source supplied either by superheated steam (800–950 °C) mixed with preheated ethylbenzene feedstock prior to exposure to the catalyst (the adiabatic process), or by indirect heat exchange design (the isothermal process). For every mole of ethylbenzene, the process produces one mole of each styrene and hydrogen. This process accounts for over 90% of the total worldwide production of styrene.

4.11.1.2.2 Halcon International's Conventional Oxidation Process

During the 1970s, Halcon International and Atlantic Richfield Chemical Company formed a joint venture company to develop a novel oxidation process to produce styrene and propylene oxide. As shown in the following reaction sequences, ethylbenzene is first oxidized to ethylbenzene hydroperoxide, which then reacts with propylene in the presence of metal oxide catalysts (such as molybdenum, tungsten, and vanadium compounds) to yield α-phenylethanol and propylene oxide. The resulting alcohol then can be dehydrated to styrene or reduced back to ethylbenzene for recycling if propylene oxide is the desired product instead of styrene. Shell also independently developed a similar dual-purpose technology to commercially produce styrene and propylene oxide. This oxidation process is currently used for as much as 10% of the total worldwide production of styrene.

4.11.2.1 Benzene-Free Synthetic Route for the Production of Styrene
4.11.2.1.1 Butadiene-to-Styrene Route

Dow Chemical has developed a two-step zeolite-based process to produce styrene from butadiene contained in crude C_4 streams. As shown in the following scheme, 1,3-butadiene (in the mixed C_4 stream) undergoes a liquid-phase cyclodimerization (Diels-Alder reaction) over a proprietary copper-loaded zeolite catalyst at moderate temperature and pressure, to give 4-vinyl-l-cyclohexene (4-VCH) with 99% selectivity. In the second step, the 4-VCH is catalytically oxidized (in the presence of steam) to styrene over one of Dow's proprietary oxide catalysts. The overall yield of styrene is greater than 90%. This process was originally tested in a 40-lb-per-hour pilot plant, and is now in commercialization.

C_4 streams are one of the by-products from the manufacture of ethylene, and contain a mixture of butadiene, butenes, and butanes. The projected growth for ethylene is much higher than that for butadiene, and a growing surplus of butadiene is expected to reach 1.36 million tons in 1998. Dow's technology apparently offers a profitable alternative for ethylene producers to upgrade crude butadiene (from C_4 streams) into more valuable styrene.

Based on Dow's estimation, this new process provides a significant economic advantage over conventional ethylbenzene-based processes. For example, even at lower production capacities, the styrene can be produced as much as 10 ¢/lb cheaper than the conventional processes. In addition, this new technology may also seem most advantageous for those countries (particularly in Asia/Pacific) that are net importers of styrene, but have an excess of butadiene (i.e., have surplus for export) due to ethylene manufacturing expansions. By applying Dow's butadiene-to-styrene process, those countries may find that it is a better economic approach than the marginal economics of existing styrene import when butadiene export prices are taken into account.

Most importantly, this alternative route for styrene production avoids the use of toxic benzene as a feedstock to produce this essential industrial chemical.

4.12 CARBON TETRACHLORIDE (CCL₄) AND TOXIC ORGANIC SOLVENTS

4.12.1.1.1 Replacement of CCl_4 and Other Toxic Organic Solvents
Substitution of toxic solvents with "safer" ones is, of course, a desirable step, but impediments to its implementation sometimes include the reaction requirements themselves. Several potential solvent substitutes exist, however, which are compatible with minimal side-product formation and good product yield. These include supercritical CO_2, 1,3-dioxolane, and DMPU, a cyclic urea.

4.12.1.1.2 Supercritical Carbon Dioxide as Replacement for CCl_4
Supercritical carbon dioxide ($SC-CO_2$) has been shown to serve as an environmentally benign substitute for a number of solvents that are typically used for free-radical reactions.[57] Such conventional reaction solvents, as benzene or the chlorinated compound, carbon tetrachloride, can be replaced successfully by $SC-CO_2$. As an example of a free-radical, side-chain bromination of all alkylaromatics using $SC-CO_2$, benzyl bromides can be prepared in high yields by direct bromination of toluene and ethylbenzene The bromine-atoms selectivity observed with $SC-CO_2$ are similar to those achieved with conventional organic solvents, and side-product formation is minimized. $SC-CO_2$ is, therefore, considered an effective alternative to CCl_4 for use in the classic Ziegler bromination reaction with the brominating reagent, N-bromosuccinimide, as shown in the following reaction scheme:

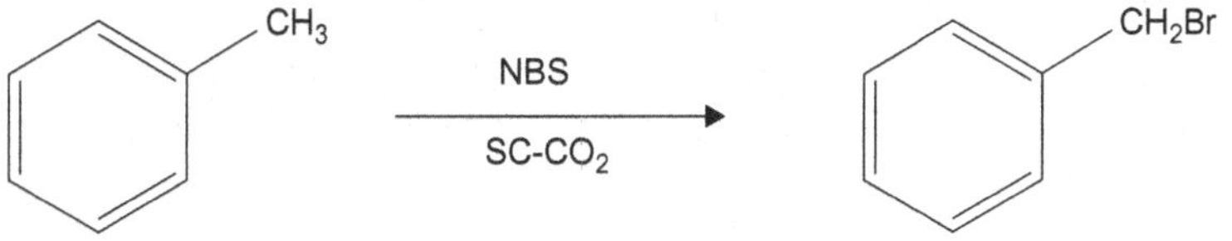

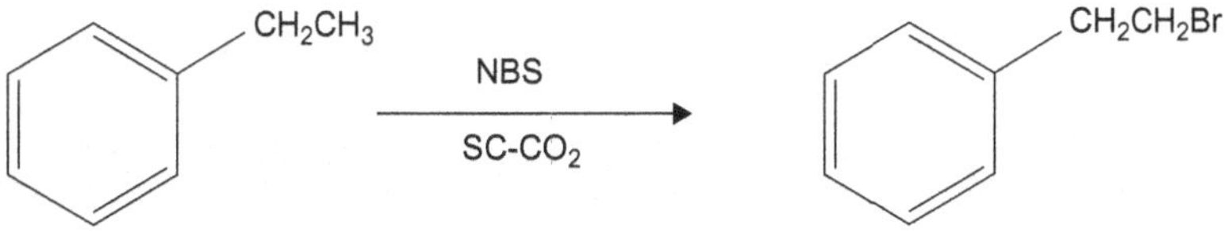

4.12.1.1.3 Dioxolane as Replacement for Other, More Toxic Organic Solvents

The cyclic diether, 1,3-dioxolane, is recommended by Ferro Corporation as a more benign solvent substitute for chlorinated organic solvents, such as methylene chloride, 1,2-dichloroethane, and 1,1,1-trichloroethane, and for ketones, such as methyl ethyl ketone (MEK). This ethylene glycol-based ether is a suitable solvent under neutral and basic conditions in several major-use areas. It is a powerful solvent for softening and dissolving polymers made from polar monomers, for example, polycarbonates, acrylates, cellulosics urethanes, phenolics, nitriles, urea-formaldehydes, and alkyds, as well as polyesters, vinyl epoxys, and halogen-containing polymers. As a reaction solvent it is added as a component to a special quaternary ammonium or phosphonium salt solution for preparation of a vesicular phenoxy resin. Other beneficial uses for the solvent dioxolane, include:

1. Extending the pot life of urea and phenol-formaldehyde resins;
2. Pre-treating polyester fibers for improved dye retention; and
3. Dissolving cross-linked alkyd resins for effective paint and film removal.

Trade-offs in using dioxolane to achieve a greener chemistry include its flammability and damaging effect on contact with some materials.

4.12.1.1.4 DMPU, a Safe Substitute for HMPA

Hexamethylphosphoric triamide (HMPA) is a colorless organic liquid with the formula [(CH3)2N]3PO. It is used as a polymer solvent, selective solvent for gases, stabilizer in polystyrene against thermal degradation, laboratory solvent for organometallic and organic reactions, and other applications. It is useful for improving the selectivity of certain organic reactions, for instance, in some deprotonation reactions it is used as an additive to break up the oligomers of lithium bases, such as butyl lithium. Also, because HMPA solvates cations so well while not solvating anions, it has been used as a solvent for some difficult SN2 reactions where an electrophilic substrate is being reacted with a nucleophile. Due to these unusual properties, it is widely used as a dipolar aprotic solvent and co-solvent in organic synthesis.

When HMPA was determined to be carcinogenic in rodent bioassays in the mid-1970s, its industrial production and use was markedly curtailed by major chemical manufacturers, including DuPont and ICI (now Zeneca); however, its use in academic and other research laboratories continues. Most common aprotic dipolar solvents, such as dimethyl sulfoxide (DMSO), nitromethane, N-methylpyrrolidone, and others, are generally unsuitable replacements for HMPA, especially with very nucleophilic and basic reagents requiring low reaction temperatures.

The cyclic urea known as DMPU (l,3-Dimethyl-3,4,5,6-tetrahydro-2(H)-pyrimidone or N,N'-dimethylpropyleneurea), has been demonstrated by Seebach and colleagues in Switzerland to closely reproduce the solvent capabilities of HMPA.[58] Moreover, with a structure quite different from HMPA, DMPU was found to be non-mutagenic and non-carcinogenic when tested to determine if it might cause similar genetic and chronic health effects and, therefore, was proposed as a safe replacement. Based on successful substitution of DMPU for HMPA, as shown in the following reaction schemes, DMPU can be recommended as a possible replacement for HMPA wherever possible in large-scale commercial processes.

1. In this Wittig olefination reaction, admixture of 50% DMPU as the cosolvent in place of HMPT achieved comparable product yield with a significantly increased shift toward the (Z)-isomer.

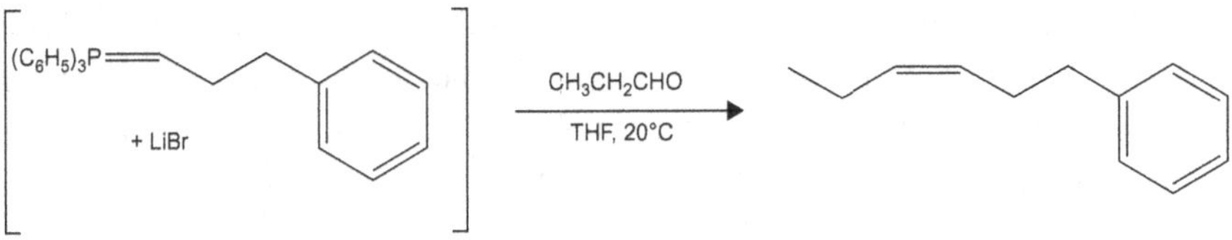

Cosolvent (% vol.)	Yield (%)	(Z)/(E)
-	46	83:17
HMPT (35)	44	92:8
DMPU (35)	39	93:7

2. In this example, the reagent lithiodithiane adds to cyclohexenone almost as effectively with DMPU as with HMPT. With the use of DMPU as replacement solvent, the 1,2-addition product (a) is suppressed in favor of the 1,4-addition product.

THF, −75°C

+

a

b

Additive (equiv.)	Yiled (%) (a+b)	a/b
-	90	2:98
2 HMPA	50–80	95:5
2 DMPU	70	82:18
4 DMPU	70	92:8

HMPA, although not an acute risk for explosions or acute toxicity, is still considered a terrorism risk since the point of terrorism is not merely to inflict immediate damage, but also to inflict fear, including the fear of the widespread risk of a cancer-causing agent.

4.12.1.1.5 Development of a Green Synthesis for Taxol® Manufacture via Plant Cell Fermentation and Extraction

Paclitaxel, the active ingredient in the anticancer drug Taxol®, was first isolated and identified from the bark of the Pacific yew tree, Taxus brevifolia, in the late 1960s. The utility of paclitaxel to treat ovarian cancer was demonstrated in clinical trials in the 1980s. The continuity of supply was not guaranteed, however, because yew bark contains only about 0.0004% paclitaxel. In addition, isolating paclitaxel required stripping the bark from the yew trees, killing them in the process. Yews take 200 years to mature, and are part of a sensitive ecosystem.

The complexity of the paclitaxel molecule makes commercial production by chemical synthesis from simple compounds impractical. Published syntheses involve about 40 steps with an overall yield of approximately 2%. In 1991, NCI signed a Collaborative Research and Development Agreement with Bristol-Myers Squibb (BMS) in which BMS agreed to ensure supply of paclitaxel from yew bark while it developed a semisynthetic route (semisynthesis) to paclitaxel from the naturally occurring compound 10-deacetylbaccatin III (10-DAB).

10-DAB contains most of the structural complexity (eight chiral centers) of the paclitaxel molecule. It is present in the European yew, Taxus baccata, at approximately 0.1% by dry weight, and can be isolated from the leaves and twigs without harm to the trees. Taxus baccata is cultivated throughout Europe, providing a renewable supply that does not adversely impact any sensitive ecosystem. However, the semi-synthetic process is complex, requiring 11 chemical transformations and seven isolations. The semi-synthetic process also presents environmental concerns, requiring 13 solvents along with 13 organic reagents and other materials.

Bristol-Myers Squibb has developed a more sustainable process using the latest plant cell fermentation (PCF) technology. PCF replaces the conventional process that extracts a paclitaxel building block from leaves and twigs of the European yew. Based on projected production and sales figures for paclitaxel, during the first 5 years of commercialization, PCF technology will eliminate an estimated 71,000 pounds of hazardous chemicals and materials, 10 solvents and 6 drying steps, and save a significant amount of energy.

In the cell fermentation stage of the process, calluses of a specific Taxus cell line are propagated in a wholly aqueous medium in large fermentation tanks under controlled conditions at ambient temperature and pressure. The feedstock for the cell growth consists of renewable nutrients: sugars, amino acids, vitamins, and trace elements. BMS now extracts paclitaxel directly from plant cell cultures, then purifies it by chromatography and isolates it by crystallization.

By replacing leaves and twigs with plant cell cultures, BMS improves the sustainability of the paclitaxel supply, allows year-round harvest, and eliminates solid biomass waste. Compared to the semi-synthesis from 10-DAB, the PCF process has no chemical transformations, thereby eliminating six intermediates. BMS is now manufacturing paclitaxel using only plant cell cultures.

Solvent use, based on the expected production volume of taxol during the first five years of using the PCF process, is shown in the Table 4.1 below.

While the overall solvent usage is higher for the new PCF process, the bulk of the organic solvent is acetonitrile used for chromatography. This material can be recovered for reuse in chromatography. It should also be considered that the main waste components of water from the PCF process are DMF and formamide. This compares favorably against the

Table 4.1 Comparative Solvent use in the Semi-Synthetic and New Plant Cell Fermentation Methods For Manufacturing Paclitaxel		
Solvent Usage	Former Semi-Synthesis Route (Liters)	New PCF Route (Liters)
Toluene	52,500	0
Isobutyl Acetate	25,200	0
Heptanes	73,500	0
Acetone	210,000	0
Methanol	27,300	0
Tetrahydrofuran	96,600	0
Methyltertbutylether	29,400	0
Ethanol	38,850	0
Ethyl Acetate	22,050	0
Glacial Acetic Acid	21,000	0
Dichloromethane	210,000	328,650
Dimethylformamide	30,450	5,775
Formamide	0	12,075
Acetonitrile	14,700	1,071,000
Isopropanol	53,550	31,500
Water	432,600	1,031,100
Total organic solvents	905,100	1,449,000

greater bio-burden in the aqueous wastes from the semi-synthetic route, which contains all the inorganic materials and amines listed above, plus trifluoroacetic acid, acetic acid, and DMF.

From the table, it can be seen that 10 solvents have been removed from processing by changing to the PCF process. In particular, the use of THF, which can form explosive peroxides, has been totally eliminated.

Other benefits from switching to the PCF process include:

1. Two cryogenic steps and three low temperature steps have been eliminated with consequent energy savings.
2. Two protection steps and two de-protection steps have been eliminated.
3. Six drying steps have been eliminated with consequent energy savings and reductions in employee exposure to dust in the workplace.

In evaluating the two processes, that is, semi-synthesis versus PCF technology, it is reasonable to compare:

1. Growing and harvesting yew bushes versus fermentation steps
2. Isolation of 10-DAB versus isolation of paclitaxel crude
3. Semi-synthesis involving seven isolations versus final chromatographic purification

The following comparison of semi-synthesis and PCF technology is based on the Twelve Green Chemistry Principles outlined by Anastas and Warner.[10]

Reagent Use

The amount of materials eliminated by using PCF technology, based on the estimated volume for the first five years of taxol production, are as follows:

1. 3.15 metric tons of hydrobenzamide
2. 2 metric tons of diisopropylethylamine
3. 1.58 metric tons of acetoxyacetylchloride
4. 3.7 metric tons of sodium hydroxide
5. 315 liters of hydrochloric acid
6. 1.58 metric tons of lipase enzyme
7. 1.05 metric tons of potassium phosphate
8. 2.94 metric tons of sodium phosphate
9. 0.11 metric tons of sodium bicarbonate
10. 0.11 metric tons of diatomaceous earth
11. 0.11 metric tons of activated carbon
12. 42 kilograms of pyridiniumparatoluenesulfonate
13. 0.74 metric tons of 2-methoxypropene
14. 32 kilograms of dimethylaminopyridine
15. 0.53 metric tons of benzoyl chloride
16. 1.89 metric tons of ammonium hydroxide
17. 0.42 metric tons of imidazole
18. 0.53 metric tons of triethylsilylchloride
19. 3.15 metric tons of lithiumhexamethyldisilazide
20. 0.21 metric tons of acetic anhydride
21. 5.78 metric tons of sodium chloride
22. 42 kilograms of butylated hydroxytoluene
23. 0.84 metric tons of trifluoroacetic acid
24. 1.26 metric tons of sodium acetate

From a life cycle perspective, consideration should also be given to the removal of environmental stresses caused by the manufacture of these materials and the waste treatment ensuing from their use in the semi-synthetic route to paclitaxel.

Aside from the obvious environmental benefits inherent in the new PCF method and its reduction in steps, solvents, reagents, and waste, the absence of these chemicals from the manufacturing process also means that they cannot become the target of a terrorist-induced accident.

4.12.1.1.6 Reduced Solvents in the Sertraline Process

Sertraline is the active ingredient in the important pharmaceutical, Zoloft®, which is the most prescribed agent of its kind and is used to treat depression, which strikes 20 million adults in the U.S. each year and costs society $43.7 billion (1990 dollars). As of 2004, more than 250 million Zoloft® prescriptions had been written in the U.S.[59]

Pfizer has dramatically improved the commercial manufacturing process of sertraline by replacing a three-step sequence in the original manufacturing process with a streamlined, single step in the new sertraline process. The new "combined" process consists of imine formation of monomethylamine with a tetralone, followed by reduction of the imine function and *in-situ* resolution of the diastereomeric salts of mandelic acid to provide chirally pure sertraline in much higher yield and with greater selectivity. A more selective palladium catalyst was implemented in the reduction step, which reduced the formation of impurities and the need for reprocessing. Raw material use was cut by 60%, 45%, and 20% for monomethylamine, tetralone, and mandelic acid, respectively.

The new combined process is further optimized by using the more benign solvent, ethanol. This change eliminated the need to use, distill, and recover four solvents (methylene chloride, tetrahydrofuran, toluene, hexane) from the original synthesis.

By using solubility differences to drive the equilibrium toward imine formation in the first reaction of the combined steps, approximately 310,000 pounds per year of the problematic reagent, titanium tetrachloride, have been eliminated. This process change eliminates 220,000 pounds of 50% sodium hydroxide, 330,000 pounds of 35% hydrochloric acid waste, and 970,000 pounds of solid titanium dioxide waste per year.

By eliminating waste, reducing solvents, and maximizing the yield of key intermediates, the innovative new sertraline process not only has significant economic and environmental advantages, but also eliminates the possibility that those hazardous materials could be the object of a hostile terrorist act.

4.12.1.1.7 Tunable Solvents Reduce Solvent Use by Coupling Reaction and Separation Processes

For any chemical process, there must be both a reaction and a separation. Generally, the same solvent is used for both, but is optimized only for the reaction. The separation typically involves 60–80% of the cost of the reaction, and almost always has a large environmental impact. Conventional reactions and separations are often designed separately, but Professors Eckert and Liotta of the Georgia Institute of Technology have optimized novel solvents for use in both reactions and separations. By replacing conventional organic solvents with their novel tunable solvents, they achieve improved reaction and separation processes and eliminate waste.

Supercritical CO_2, near-critical water, and CO_2-expanded liquids are tunable benign solvents that offer exceptional opportunities as replacements for hazardous solvents. They generally exhibit better solvent properties than gases and better transport properties than liquids. They offer substantial property changes with small variations in thermodynamic conditions, such as temperature, pressure, and composition. They also provide wide-ranging environmental advantages, from human health to pollution prevention and waste minimization.

These researchers have used supercritical CO_2 to tune reaction equilibria and rates, improve selectivity, and eliminate waste. They were the first to use supercritical CO_2 with phase transfer catalysts to separate products cleanly and economically. Their method allows them to recycle their catalysts effectively. They have demonstrated the feasibility of a variety of phase transfer catalysts on reactions of importance in the chemical and pharmaceutical industries, including chiral syntheses. They have carried out a wide variety of synthetic reactions in near-critical water, replacing conventional organic solvents. This includes acid- and base-catalysis using the enhanced dissociation of near-critical water, negating the need for any added acid or base, and eliminating subsequent neutralization and salt disposal. They have used CO_2 to expand organic fluids to make it easier to recycle homogeneous catalysts, including phase transfer catalysts, chiral catalysts, and enzymes. Finally, they have used tunable benign solvents to design syntheses that minimize waste by recycling, and demonstrate commercial feasibility by process economics.

The team of Eckert and Liotta has combined state-of-the-art chemistry with engineering know-how, generating support from industrial sponsors to facilitate technology transfer. They have worked with a wide variety of government and industrial partners to identify the environmental and commercial opportunities available with these novel solvents; their interactions have facilitated the technology transfer necessary to implement their advances.

Combining reactions with separations in a synergistic manner, this method uses benign solvents, minimizes waste, and improves performance, and the indirect result is that companies choosing to employ the method would lower their exposure to terrorists seeking to target facilities holding large quantities of hazardous chemical materials.

4.12.1.1.8 New Enzyme Technology to Improve Paper Recycling
Paper mills traditionally use hazardous solvents, such as mineral spirits, to remove sticky contaminants, referred to as "stickies," from machinery. A new technology developed by Buckman Laboratories International, called Optimyze®, uses a novel enzyme to remove stickies from paper products prior to recycling, increasing the percentage of paper that can be recycled. In one U.S. mill, conversion to Optimyze® reduced solvent use by 200 gallons per day and chemical use by about 600,000 pounds per year. Production increased by more than 6%, which amounted to a $1 million benefit per year for this mill alone. A major component of the sticky contaminants in paper is poly (vinyl acetate) and similar materials. Optimyze® contains an esterase enzyme that catalyzes the hydrolysis of this type of polymer to poly (vinyl alcohol), which is not sticky, and is water-soluble. A bacterial species produces large amounts of the Optimyze® enzyme by fermentation. As a protein, the enzyme is completely biodegradable, much less toxic than alternatives, and much safer. Furthermore, only renewable resources are required to manufacture it. It has been commercially available since May 2002, and in that short time, more than 40 paper mills have converted to Optimyze® for manufacturing paper goods from recycled papers.

Table 4.2 Health and Safety Considerations Of Conventional vs Enzymatic Methods for Removing Stickies

shwarya	Mineral Spirits	Enzymatic Cleaner (Optimyze)
Source	Petroleum hydrocarbons	Renewable materials (fermentation)
Flammability	Flammable	Water-based, not combustible
Flashpoint	110°(44 °C)	No flashpoint
Worker effects	Odors, irritation	Not noticeable
Toxicity	Can be fatal (ingestion, inhalation)	Slight irritation possible when undiluted
VOC content	100%	15%
HAPs	7–8%	None detected
Aquatic toxicity	10–20 ppm (rainbow trout)	700 ppm (zebra fish)
Ecotoxicity	EC50 estimated at 1–10 ppm	No effect at >1000 ppm

In the production of paper from recycled materials, this enzyme eliminates the need for using hydrocarbon solvents or similar chemicals that are more toxic, more flammable, and derived from non-renewable petroleum-based feedstocks. Paper mills adopting Optimyze® have been able to greatly reduce the use of hazardous, flammable solvents.

A side-by-side comparison of the conventional and greener methods illustrates the relatively greater hazards to be confronted if a hostile strike were to provoke uncontrolled release from the conventional factory that relies on mixed solvents (Table 4.2).

4.12.1.1.9 Reducing Solvents and Reagent Requirements Through Activation of C-H Bonds

The preparation of high-valued organic chemicals often involves lengthy, multi-step synthetic sequences. These typically require large amounts of various chemical reagents, such as oxidizing and reducing agents, drying agents, and organic solvents for the performance of the reactions. Large amounts of organic solvent are also often required for the separation of the desired products from one another, especially when chromatography is employed. Use of these reagents and solvents entails serious risks of accident and injury.

One effective way to address this problem is to drastically reduce the number of chemical steps required in these synthetic sequences. University of California at Berkeley's Prof. Bergman has demonstrated that the employment of C-H activation reactions within synthetic sequences provides important progress toward this goal. The Bergman group has pioneered the direct activation of C-H bonds in organic molecules that are found in locations remote from other functional groups. These C-H activation reactions are now being used successfully in the synthesis of various chemicals and pharmaceutical products. Ultimately, this should have a profound impact on various fields and sectors of chemical manufacturing and production. Studies of the mechanism of C-H activation have also provided a substantial amount of fundamental information about this important process, such as the factors that promote high activation of different types of C-H bonds in hydrocarbons.

4.12.1.1.10 Biocatalyst Reduces Solvent and Heavy Metal Needs in Pharmaceutical Manufacturing

The synthesis of an active drug ingredient is frequently accompanied by the generation of a large amount of waste, as numerous steps are commonly necessary, each of which may require feedstocks, reagents, solvents, and separation agents. Lilly Research Laboratories has developed a novel process that decreases the waste generated during the synthesis of drugs, such as the anticonvulsant drug candidate, LY300164, a pharmaceutical agent being developed for the treatment of epilepsy and neurodegenerative disorders.

The new process eliminates approximately 41 gallons of solvent and 3 pounds of chromium waste for every pound of LY300164. It also improved worker safety and increased its product yield from 16% to 55%.

The original synthesis used to support clinical development of the drug candidate proved to be an economically viable process, although several steps proved problematic. A large amount of chromium waste was generated, an additional activation step was required, and the overall process required a large volume of solvent.

The new synthesis begins with the biocatalytic reduction of a ketone to an optically pure alcohol. The yeast *Zygosaccharomyces rouxii* demonstrated good reductase activity, but was sensitive to high product concentrations. To circumvent this problem, a novel three-phase reaction design was employed. The starting ketone was charged to an aqueous slurry containing a polymeric resin, buffer, and glucose, with most of the ketone adsorbed on the surface of the resin. The yeast reacted with the equilibrium concentration of ketone remaining in the aqueous phase. The resulting product was adsorbed onto the surface of the resin, simplifying product recovery. All of the organic reaction components were removed from the aqueous waste stream, permitting the use of conventional wastewater treatments.

A second key step in the synthesis was selective oxidation to eliminate the unproductive redox cycle present in the original route. The reaction was carried out using dimethylsulfoxide, sodium hydroxide, and compressed air, eliminating the use of chromium oxide (a possible carcinogen) and preventing the generation of chromium waste. The new protocol was developed by combining innovations from chemistry, microbiology, and engineering. Minimizing the number of changes to the oxidation state improved the efficiency of the process while reducing the amount of waste generated. The alternative synthesis presents a novel strategy for producing 5H-2,3-benzodiazepines. The approach is general, and has been applied to the production of other anticonvulsant drug candidates. The technology is low-cost and easily implemented, and should have broad applications within the manufacturing sector.

Upon implementing the new synthetic strategy, roughly 9000 gallons of solvent and 660 pounds of chromium waste were eliminated for every 220 pounds of LY300164 produced. Only three of the six intermediates generated were isolated, limiting worker exposure and decreasing processing costs. The synthetic scheme proved more efficient as well, with percent yield climbing from 16% to 55%.

By drastically reducing the amount of solvent and heavy metals required to synthesize LY300164, Lilly Labs has also reduced the amounts of those hazardous substances needed on hand at the production site as well as any consequences pursuant to an accident or act of terrorism aimed at that manufacturing facility.

4.12.1.1.11 Membrane-Based Process for Producing Lactate Esters Promises Non-Toxic Alternative to Halogenated and Toxic Solvents

Argonne National Library (ANL) has developed a novel process, referred to as the Argonne process, based on selective membranes that permit low-cost synthesis of high-purity ethyl lactate and other lactate esters from carbohydrate feedstock. These esters can replace a wide variety of volatile organic compounds (VOCs) used as solvents. The process requires little energy, is highly efficient and selective, eliminates the large volumes of salt waste generated by conventional processes, and reduces pollution and emissions. Lactate esters can potentially replace 7.6 billion pounds of toxic solvents used annually by industry, commerce, and households.

ANL's novel process uses pervaporation membranes and catalysts. In the process, ammonium lactate is thermally and catalytically cracked to produce the acid, which with the addition of alcohol is converted to the ester. The selective membranes pass the ammonia and water with high efficiency while retaining the alcohol, acid, and ester. The ammonia is recovered and reused in the fermentation to make ammonium lactate, eliminating the formation of waste salt. The innovation overcomes major technical hurdles that had made current production processes for lactate esters technically and economically noncompetitive. The innovation will enable the replacement of toxic solvents widely used by industry and consumers, expand the use of renewable carbohydrate feedstocks, and reduce pollution and emissions.

Ethyl lactate has a good temperature performance range (boiling point: 309 °F, melting point: 104 °F), is compatible with both aqueous and organic systems, is easily biodegradable, and has been approved for food by the U.S. Food and Drug Administration (FDA). Lactate esters (primarily ethyl lactate) can replace most halogenated solvents (including ozone-depleting chlorofluorocarbons (CFCs), carcinogenic methylene chloride, toxic ethylene glycol ethers, perchloroethylene, and chloroform) on a 1:1 basis. At 1998 prices ($1.60–2.00 per pound), the market for ethyl lactate is about 20 million pounds per year for a wide variety of specialty applications. The novel and efficient ANL membrane process will reduce the selling price of ethyl lactate to $0.85–1.00 per pound and enable it to compete directly with the petroleum-derived toxic solvents currently in use. The favorable economics of the ANL membrane process, therefore, can lead to the widespread substitution of petroleum-derived toxic solvents by ethyl lactate in electronics manufacturing, paints and coatings, textiles, cleaners and degreasers, adhesives, printing, de-inking, and many other industrial, commercial, and household applications.

More than 80% of the applications—requiring the use of over 7.6 billion pounds of solvents—in the United States each year are suitable for reformulation with environmentally friendly lactate esters.

The ANL process has been patented for producing esters from all fermentation-derived organic acids and their salts. Organic acids and their esters, at the purity achieved by this process, offer great potential as intermediates for synthesizing polymers, biodegradable plastics, oxygenated chemicals (e.g., propylene glycol and acrylic acid), and specialty products. By improving purity and lowering costs, the ANL process promises to make fermentation-derived organic acids an economically viable alternative to many chemicals and products currently derived from petroleum feedstocks.

A U.S. patent on this technology has been allowed, international patents have been filed, and NTEC, Inc. has licensed the technology for lactate esters and provided the resources for a pilot-scale demonstration of the integrated process at ANL. The pilot-scale demonstration has produced a high-purity ethyl lactate product that meets or exceeds all the process performance objectives. A 100-million-pound-per-year full-scale plant is planned.

It is interesting to note that this effort did not win for creating a new solvent. Their product, ethyl lactate, has been known for years as a technically effective alternative solvent, and approved by the FDA for use in food. Until this innovation, ethyl lactate has been too expensive to employ as an alternative solvent. ANL's breakthrough transformed the economics of producing ethyl lactate that allows it to compete with existing solvents. The ANL team's primary breakthrough was developing a cost-cutting manufacturing process based on new membrane technology that enabled more cost effective separation and purification techniques. It is instructive to review how ANL performed it in a fashion that is more environmentally benign and illustrates most of the 12 Principles of Green Chemistry[60]:

1. The process eliminates salt waste (gypsum) and undesirable by-products achieving Principle 1, Prevent Waste.
2. The process innovation demonstrates Principle 2, Atom Economy in that undesirable by-products are avoided, much more of the input materials are incorporated in the final product, and
3. ANL's synthetic process minimizes the use of ethanol and the need to distill it and keeps the alcohol in the reaction system reducing the risk of fire or explosion—fulfilling Principle 3, Less Hazardous Synthesis.

4. Ethyl Lactate is non-toxic and therefore a good example of Principle 4, Safer Chemicals.
5. Their innovation provides an option for users to address Principle 5, Alternative Solvents.
6. The use of catalysis and the membrane separations technology enables the reaction to consume 90% less energy than traditional processes fulfilling Principle 6, Energy Efficiency.
7. This process for ethyl lactate is carbohydrate-based, rather than petrochemical-based, using corn, for example. It therefore demonstrates Principle 7, Renewable Feedstocks.
8. The process illustrates Principle 9 in its use of catalysis for cracking the carbohydrates.
9. Ethyl lactate is biodegradable. It hydrolyses into ethyl alcohol and lactic acid—common constituents of food. Thus, the end product builds in Principle 10, Design for Degradation.
10. The handling of the end product is less dangerous than most other solvents. Therefore it illustrates Principle 12, Accident Prevention.[60]

By taking these steps towards a green chemistry production model, Argonne and the licensing manufacturers have also dramatically reduced their facilities' vulnerability to accidents or deliberate acts against them.

4.13 A SAFER GRIGNARD REAGENT

Grignard reagents can prove hazardous because they become explosive and may burst into flames on contact with moist air. According to Ferro Corporation, Grignard reagents made with the solvent, diethylene glycol dibutyl ether (butyl diglyme), are safer and more environmentally friendly when used in industrial chemical processes as substitutes for Grignards prepared with other solvents. Diglymes are saturated polyethers with no other functional groups, and, therefore, act as relatively inert aprotic polar solvents. Because it has a much higher flash point and boiling point than either tetrahydrofuran (THF) or diethyl ether (DEE), using butyl diglyme as a replacement for either of these other two common solvents produces safer Grignard reagents. For example, methyl magnesium chloride (CH_3MgCl) can be prepared directly in 2.5 molar butyl diglyme, producing a true, clear, single-phase Grignard solution.

4.14 LOWER VOC COATINGS

Background: Volatile organic compounds (VOCs) are organic chemical compounds that have high enough vapor pressures under normal conditions to vaporize and enter the atmosphere, and encompass a wide range of carbon-based molecules, such as aldehydes, ketones, and hydrocarbons, including methane, benzene, xylene, and 1,3-butadiene, the first two of which are given individual treatment in this report (see sections on methane conversion and benzene).

Common artificial sources of VOCs include paint thinners, dry cleaning solvents, and some constituents of petroleum fuels (e.g., gasoline and natural gas). When accidentally released into the environment, they can become soil and groundwater contaminants, and their vapors escaping into the air contribute significantly to air pollution and global warming.

In response to environmental concerns, efforts are on-going under both federal and state auspices to reduce levels of volatile organic chemicals (VOCs) in industrial paints. The volatile organic solvent content of paints is measured as pounds of solvent per gallon of paint. From levels of greater than 5 lbs/gal common before 1970, current legislation is moving toward limits of VOC content to levels of 3 lbs/gal and below.[8]

Process design changes that reduce or eliminate the conventional VOCs in paint and coating materials are helping to achieve the goal of substantially decreasing airborne pollutants that contribute to ecological harm and respiratory illnesses. As a side-benefit, some of the replacement technologies also lower the industry's susceptibility to terrorist attack by eliminating some of the most hazardous, explosive constituents and feedstocks.

Besides VOC content reduction in coating formulations, other approaches are being used to achieve lower emissions in compliance with clean-air regulations. These include improving the efficiency of application and transfer methods and developing new technologies that are environmentally friendlier. Coating application methods fall under a variety of regulations which have encouraged the development of alternative techniques with greater transfer efficiency, as well as the use of add-on control devices. Some of the new technologies that achieve lowered chemical demand include high-solids, waterborne, powder, UV-cure, and two-component (catalyzed) coating systems.

4.14.1.1 Conventional Coating Utilization of VOCs

Industrial and special purpose coatings technologies have traditionally been solvent-based. Several types of these conventional coatings, which contain volatile organic compounds (VOCs) as solvents, are the following: conventional low-solids, lacquers, two-part catalyzed, high-solids, and non-aqueous dispersions. Environmental constraints resulting in efforts to reduce emissions of VOCs from conventional coatings have been driving the coatings sector of the chemical industry to adopt alternative coatings technologies that reduce or eliminate VOC usage.

4.14.1.2 Alternative Approaches to Coatings

4.14.1.2.1 Lower VOC Powder Coatings

Dramatic growth in the use of powder coatings has been fostered by its promotion as a virtually pollution-free method of applying finishing coats.[61] The global powder coating market saw a demand of 1680.0 kilo tons in 2011. By 2018, the demand is projected to rise to 2667.6 kilo tons, showing a 6.8% Compound Annual Growth Rate (CAGR). In revenue terms, the global powder coatings market stood at US$6.5 billion in 2011, and will register a healthy CAGR of 7.2% through the forecast period (2012−2018) to notch US$10.5 billion by 2018.[62] The coating material is in the form of a fine, dry powder containing no VOCs. Thermosetting powder coatings used in metal finishing operations require curing at temperatures in the range of 300−400 °F.

Amoco Chemicals has a 100% solids polyester coating used as a durable finish for outdoor furniture. It is based on their Cargill 3000 polyester resin, which is synthesized from terephthalic acid as the primary ingredient. The Belgian company, Solvay, has developed polyvinylidene fluoride polymers for electrostatic application to metal surfaces. The product line is Solvay's PVDF Solef series. The electrostatically charged particles are applied with pneumatic force to an electrically conductive article or surface. According to Amoco, the environmental benefit of completely eliminating solvent evaporation into the atmosphere is coupled with the achievement of more effective surface coverage and savings in both volumes of materials used and cost to the company.

4.14.1.2.2 Lower VOC Radiation-Cured High Solids Coatings

Radiation-curable coatings include high-solids materials that are formulated to be cured by ultraviolet (UV) and electron beam (EB) techniques. While their current use is at a much lower level than powder coatings, and the growth rate is slower, radiation curable coatings have some major advantages. They are environmentally favorable in that their constituents are 100% solids with reduced emissions, faster curing speed, and lowered cure temperatures not requiring the provision of thermal energy sources such as ovens.

Research challenges include improving coating quality and identifying less toxic monomers than the typically utilized unsaturated monomers that polymerize when UV light is applied in the presence of a photoinitiator. Conventional systems have been based on the copolymerization of unsaturated polyesters and vinyl monomers, or on polyfunctional acrylates. A promising newer system has been described that utilizes vinyl ether monomers cationically cured with epoxies, and anticipates expanding applications. Although there are trade-offs associated with adaptation of coatings processes to radiation curing, these more benign room-temperature curing mechanisms provide an overall benefit to the environment, which favors their continued development and use.

4.14.1.2.3 Lower VOC Coatings Via Microwave Irradiation

In the search to design industrial chemical processes with gentler conditions, the utilization of microwave irradiation is growing as a means of applying heat without the use of solvents. For example:

Microwave drying—The pharmaceutical manufacturer, GlaxoSmith-Kline, has reported a new approach to solids processing.[63] It provides an alternative to the conventional production technology, which relies on high-shear granulation and fluid-bed drying. The alternative approach is described as a single-pot system with microwave/vacuum drying, and this process provides the advantage of rapid drying at a lower temperature. In addition, this novel adaptation incorporates microwave technology into an already available single-pot process design with the added benefits of shorter overall process time and fewer material-handling steps. Besides the benefits to the environment, the process offers distinct benefits to the manufacturer over the conventional alcoholic-granulation process, such as batch size equivalency, savings in energy use and equipment expenditures, and reduced demand for environmentally controlled floor space needed to process formulations.

Microwave heat source as a reaction condition—Another novel process design was reported by Dagani (1993).[64] Ordinary microwave irradiation for as little as three minutes produced chalcopyrite-type semiconductor materials from stoichiometric mixtures of metal and chalcogen powders. The solid state reactions can be accomplished with safe microwave heating to temperatures above 1000 °C. The conventional method for making bulk quantities of these semiconductors involves lengthy heating for approximately 24 h inside a furnace. While research is on-going to achieve product of sufficiently high quality for semiconductor use, the microwave technique is being offered as a new, rapid way to make a wide variety of ternary and quaternary chalcogenide compounds in bulk.

4.14.1.2.4 UV-Curable, One-Component, Low-VOC Automobile Refinish Primer

BASF has developed an auto paint primer that contains only 1.7 pounds of volatile organic compounds (VOCs) per gallon, in contrast to 3.5–4.8 pounds of VOCs per gallon of conventional primers, a reduction of over 50%. It also does not contain any diisocyanates, a major source of occupational asthma. Use in repair facilities has shown that only one-third as much of this primer is needed compared to conventional primer, and that waste is reduced from 20% to nearly zero.

The market for automotive refinish coatings in North America exceeds $2 billion for both collision repairs and commercial vehicle applications. Over 50,000 body shops in North America use these products. For more than a decade, automotive refinishers and coating manufacturers have dealt with increasing regulation of emissions of volatile organic compounds (VOCs). At first, coating manufacturers were able to meet VOC maximums with high-performance products such as two-component reactive urethanes, which require solvents as carriers for their high-molecular-weight resins. However, as thresholds for VOCs became lower, manufacturers had to reformulate their reactive coatings, and the resulting reformulations were slow to set a film. Waterborne coatings are also available, but their utility has been limited by the time it takes the water to evaporate. Continuing market pressures demanded faster film setting, without compromising either quality or emissions.

BASF's new urethane acrylate oligomer primer system uses a resin that cross-links with monomers (added to reduce viscosity) into a film when the acrylate double bonds are broken by radical propagation. The oligomers and monomers react into the film's cross-linked structure, improving adhesion, water resistance, solvent resistance, hardness, flexibility, and cure speed. The primer cures in minutes by visible or near-ultraviolet (UV) light from inexpensive UV-A lamps or even sunlight. As a UV-cured primer, it eliminates the need for bake ovens that cure the current primers, greatly reducing energy consumption. BASF claims that this primer performs better than the current conventional urethane technologies in a variety of ways: it cures 10 times faster, requires fewer preparation steps, has a lower application rate, is more durable, controls corrosion better, and has an unlimited shelf life. BASF is currently offering its UV-cured primers in its R-M® line as Flash Fill™ VP126 and in its Glasurit® line as 151-70.

Because BASF's primer contains more than 50% fewer VOCs than conventional primers, it meets even the stringent requirements of South Coast California. The one-component nature of the product reduces hazardous waste and cleaning of equipment, which typically requires solvents. The BASF acrylate-based technology requires less complex and less costly personal protective equipment (PPE) than the traditional isocyanate-based coatings; this, in turn, increases the probability that small body shops will purchase and use the PPE, increasing worker safety. To fully support its claims, BASF has conducted an eco-efficiency study with an independent evaluation.

BASF plans to offer this eco-efficient product as the first part of a complete automobile refinishing coating system, followed by their globally available waterborne basecoat (sold under the Glasurit® brand as line 90), and finished with the application of their one-component, UV-A-curable clearcoat.

The BASF refinishing system seems to offer quality, energy efficiency, economy, and speed for the small businessman operating a local body shop, while respecting the health and safety of the workers in this establishment and the environment in which these products are manufactured and used. The lower VOCs and almost zero-waste generated in production of this suite of products makes the BASF facility a less attractive target for those seeking to cause accidents with catastrophic consequences.

4.14.1.2.5 Nonvolatile Coalescent for Reduction of VOCs in Latex Paints

Coalescents are critical to the performance of latex coatings. This new coalescent eliminates the emissions of volatile organic compounds

(VOCs) associated with traditional coalescents, while performing the same function. This biobased coalescent has other advantages as well, such as lower odor, increased scrub resistance, and better opacity.

Since the 1980s, waterborne latex coatings have found increasingly broad acceptance in architectural and industrial applications. Traditional latex coatings are based on small-particle emulsions of a synthetic resin, such as acrylate- and styrene-based polymers. They require substantial quantities of a coalescent to facilitate the formation of a coating film as water evaporates after the coating is applied. The coalescent softens (plasticizes) the latex particles, allowing them to flow together to form a continuous film with optimal performance properties. After film formation, traditional coalescents slowly diffuse out of the film into the atmosphere. The glass transition temperature of the latex polymer increases as the coalescent molecules evaporate, and the film hardens. Alcohol esters and ether alcohols, such as ethylene glycol monobutyl ether (EGBE) and Texanol® (2,2,4-trimethyl-1,3-pentanediol monoisobutyrate), are commonly used as coalescents. They are also volatile organic compounds (VOCs). Both environmental concerns and economics continue to drive the trend to reduce the VOCs in coating formulations. Inventing new latex polymers that do not require a coalescent is another option, but these polymers often produce soft films and are expensive to synthesize, test, and commercialize. Without a coalescent, the latex coating may crack and may not adhere to the substrate surface when dry at ambient temperatures.

A new coalescent from Archer Daniels Midland Co., called Archer RC™, provides the same function as traditional coalescing agents, but eliminates the unwanted VOC emissions. Instead of evaporating into the air, the unsaturated fatty acid component of the product oxidizes and even cross-links into the coating. Archer RC™ is produced by inter-esterifying vegetable oil fatty acid esters with propylene glycol to make the propylene glycol monoesters of the fatty acids. Corn and sunflower oils are preferred feedstocks because they have a high level of unsaturated fatty acids and tend to resist the yellowing associated with linolenic acid, found at higher levels in soybean and linseed oils. Because Archer RC™ remains in the coating after film formation, it adds to the overall solids of a latex paint, providing an economic advantage over volatile coalescents.

The largest commercial category for latex paint, the architectural market, used 618 million gallons in the United States in 2001. Coalescing solvents typically constitute 2–3% of the finished paint by volume, corresponding to an estimated 120 million pounds of coalescing solvents in the United States, and perhaps three times that amount globally. Currently, nearly all of these solvents are lost into the atmosphere each year.

Archer RC™ has been tested in a number of paint formulations in place of conventional coalescing solvents and, in these tests, it performed as well as commercial coalescents, such as Texanol®. It often had other advantages as well, such as lower odor, increased scrub resistance, and better opacity. Paint companies and other raw material suppliers have demonstrated success formulating paints with Archer RC™ and their existing commercial polymers.

If this product, which has been commercially available since March 2004, were to be widely adopted as a nonvolatile alternative to the hundreds of millions of pounds of volatile coalescing solvents currently present in paints, the environmental, health, and security implications would be enormous. Paint manufacturing facilities that no longer used or stored large quantities of alcohol esters and ether alcohols would no longer constitute such an explosive or hazardous focal point in their communities.

4.14.1.2.6 Waterborne Polyurethane Coatings Reduce VOCs by 50%
New polyurethane coatings from Bayer use water to replace most or all of the organic solvents used in conventional two-component (2K), solvent-borne polyurethane coatings. These 2K waterborne polyurethane coatings reduce volatile organic compound (VOC) emissions by 50–90% and hazardous air pollutant (HAP) emissions by 50–99%.

Two-component (2K) waterborne polyurethane coatings are an outstanding example of the use of alternative reaction conditions for green chemistry. This technology is achieved by replacing most or all of the volatile organic compounds (VOCs) and hazardous air pollutants (HAPs) used in conventional 2K solvent-borne polyurethane coatings with water as the carrier, without significant reduction in performance of the resulting coatings. This may seem an obvious substitution, but due to the particular chemistry of the reactive components of polyurethane, it is not that straightforward.

Two-component solvent-borne polyurethane coatings have long been considered in many application areas to be the benchmark for high-performance coatings systems. The attributes that make these systems so attractive are fast curing under ambient or bake conditions, high-gloss and mirror-like finishes, hardness or flexibility as desired, chemical and solvent resistance, and excellent weathering. However, the traditional carrier has been an organic solvent that, upon cure, is freed to the atmosphere as VOC and HAP material. High-solids systems and aqueous polyurethane dispersions ameliorate this problem, but do not go far enough.

An obvious solution to the deficiencies of 2K solvent-borne polyurethanes and aqueous polyurethane dispersions is a reactive 2K polyurethane system with water as the carrier. In order to bring 2K waterborne polyurethane coatings to the U.S. market, new waterborne and water-reducible resins had to be developed. To overcome some application difficulties, new mixing/spraying equipment was also developed. For the technology to be commercially viable, an undesired reaction of a polyisocyanate cross-linker with water had to be addressed, as well as problems with the chemical and film appearance resulting from this side reaction. The successful invention of 2K waterborne polyurethanes has resulted in a technology that will provide several health and environmental benefits. VOCs will be reduced by 50–90% and HAPs by 50–99%. The amount of chemical byproducts evolved from films in interior applications will also be reduced, and rugged interior coatings with no solvent smell will consequently be available.

2K waterborne polyurethane is being applied on industrial lines where good properties and fast cure rates are required for such varied products as metal containers and shelving, sporting equipment, metal- and fiberglass-reinforced utility poles, agricultural equipment, and paper products. In flooring coatings applications where the market driving force is elimination of solvent odor, 2K waterborne polyurethane floor coatings provide a quick dry, high resistance to abrasion and lack of solvent smell (<0.1 pound per gallon organic solvent). In wood applications, 2K waterborne polyurethane coatings meet the high-performance wood finishes requirements for kitchen cabinet, office, and laboratory furniture manufacturers, while releasing minimal organic solvents in the workplace or to the atmosphere. In the United States, the greatest market acceptance of 2K waterborne polyurethane is in the area of special-effect coatings in automotive applications. These coatings provide the soft, luxurious look and feel of leather to hard plastic interior automobile surfaces, such as instrument panels and air bag covers. Finally, in military applications, 2K waterborne polyurethane coatings are being selected because they meet the demanding military performance criteria that include flat coatings with camouflage requirements, corrosion protection, chemical and chemical agent protection, flexibility, and exterior durability. The dramatic VOC reductions made possible by this technology will make the coatings manufacturing facilities safer.

4.14.1.2.7 Environmentally Responsible Fire Extinguishment and Cooling Agent

PYROCOOL Technologies, Inc. has developed a fire extinguishing foam, called PYROCOOL Fire Extinguishing Foam (PFEF) as a replacement for ozone-depleting halon gases and aqueous film-forming foams, which release both toxic hydrofluoric acid and fluorocarbons into the environment during use. PFEF is a nontoxic blend of highly biodegradable surfactants designed for use in very small quantities as a universal fire extinguishment and cooling agent that is effective at approximately one-tenth the concentration of conventional fire extinguishing chemicals.

Advances in chemical technology have greatly benefited firefighting in this century. From the limitation of having only local water supplies at their disposal, firefighters have been presented, over the years, with a wide variety of chemical agents as additives or alternatives to water to assist them. However, in actual use, these advances in chemical extinguishment agents have themselves created certain long-term environmental and health problems that can outweigh their firefighting benefits.

Halon gases, hailed as a tremendous advance when first introduced, have since proven to be extremely destructive to the ozone layer, having an ozone depletion potential (ODP) value of 10−16 times that of common refrigerants. Aqueous film-forming foams (developed by the U.S. Navy in the 1960s to combat pooled-surface, volatile, hydrocarbon fires) release both toxic hydrofluoric acid and fluorocarbons when used. The fluorosurfactant compounds that make these agents so effective against certain types of fires render them resistant to microbial degradation, often leading to contamination of ground water supplies and failure of wastewater treatment systems.

In contrast to preexisting fire retardants, PFEF is formulated to contain no glycol ethers or fluorosurfactants, instead employing biodegradable nonionic surfactants, anionic surfactants, and amphoteric surfactants with a very low mixing ratio (with water) of 0.4%. Tests show PFEF is effective against a broad range of combustibles, and in practice, it carries the distinction of extinguishing a large oil tanker fire at sea (a fire estimated by Lloyd's of London to require 10 days to extinguish) on board the Nassia tanker in the Bosporus Straits in just 12.5 minutes, saving 80% of the ship's cargo and preventing 160 million pounds of crude oil from spilling into the sea.

Its selective employment of rapidly biodegradable substances dramatically enhances the effectiveness of simple water, while eliminating the environmental and toxic impact of other traditional fire extinguishment agents. Because PFEF is mixed with water at only 0.4%, an 87–93% reduction in product use is realized, compared to conventional extinguishment agents typically used at 3–6%. Fire affects all elements of industry and society and no one is immune from its dangers. PFEF provides an innovative, highly effective, and green alternative for firefighters and eliminates the significant hazards associated with manufacture and use of previous fire-fighting agents and their hydrofluoric acid by-products.

Industries whose manufacturing relies heavily on use of VOCs constitute an attractive target to terrorists for the simple fact that, by definition, VOCs have a low flash point and, therefore, tend to be explosive. Consequently, as companies can shift their production models towards chemistry with fewer VOCs, they will also be helping to reduce their likelihood of being targeted by terrorists for an attack.

4.15 PESTICIDES

Background: Conventional pesticides such as organophosphates and carbamates, act broadly to disrupt an animal's nervous system and are, therefore, effective against a broad range of target pests, but are also toxic to humans, other mammals, and beneficial insects. Exposure to insecticides may cause acute symptoms of headache, diarrhea, dizziness, blurred vision, weakness, nausea, cramps, chest tightening, nervousness, sweating, pinpoint pupils, tearing, salivation, pulmonary edema, muscle twitching, convulsions, coma, loss of reflexes, and sphincter control.

Certainly the most catastrophic accident involving pesticides was the 1984 disaster at the Union Carbide pesticide factory in Bhopal, India, which is estimated to have killed approximately 15,000 people when a relief valve lifted on a storage tank containing methyl isocyanate, releasing a cloud of toxic gas onto residential areas surrounding the plant.[22]

Even relatively low-toxicity pesticides can have powerful consequences if released in high concentrations. For example, an accident at a pesticide factory in Mexico in 2000 occurred when emergency valves opened automatically after pressure rose in tanks at the plant holding malathion, sending 120 people to the hospital and forcing evacuation of many hundreds more.[65] Malathion is a relatively mild pesticide, and is not considered to pose a serious risk to people, according to U.S. Centers for Disease Control, yet it was also suspected of sickening 123 people in Florida in 1999. Accidents—or terrorist attacks—involving more hazardous pesticides can obviously wreak much more catastrophic consequences.

Advances in more recent pesticide research has resulted in certain insecticidal compounds that provide narrow-spectrum pest control that is safe for humans and other mammals and poses little threat to non-target species.[66] Because modern agriculture involves such large volumes of chemical inputs, spread over such vast territories, green chemistry applied to crop production—and especially to pest management—can greatly reduce the number of viable agrochemical targets available to terrorists.

4.15.1.1.1 Novel Pest Control Agent Acts by Fortifying Plants' Own Defense Mechanisms

EDEN Bioscience Corporation has developed a new class of nontoxic, naturally occurring, biodegradable proteins as an alternative to pesticides. The newly discovered proteins, called harpins, function by activating a plant's existing defense and growth mechanisms without altering the plant's DNA, thereby increasing crop yield and quality, and minimizing crop losses. When applied to crops, harpins increase plant biomass, photosynthesis, nutrient uptake, and root development.

Sold under the name Messenger®, the harpin-containing product is commercially available and is manufactured using a water-based fermentation system that uses no harsh solvents or reagents, requires only modest energy inputs, and generates no hazardous chemical wastes. Fermentation byproducts are fully biodegradable and safely disposable, and 70% of the dried finished product consists of an innocuous food grade substance that is used as a carrier for the harpin protein.

The Food and Agriculture Organization of the United Nations estimates that annual losses to growers from pests reach $300 billion worldwide. Growers in modern agricultural systems have relied heavily on either chemical pesticides or crops that are genetically engineered for pest resistance as ways to limit their economic losses and increase yields. Each of these approaches has come under criticism from environmental groups, government regulators, consumers, and labor advocacy groups due to lasting impacts on the environment.

Messenger® has been demonstrated on more than 40 crops to effectively stimulate plants to defend themselves against a broad spectrum of viral, fungal, and bacterial diseases, including some for which there currently is no effective treatment. In addition, it has been shown through a safety evaluation to have virtually no adverse effect on any of the non-target organisms tested, including mammals, birds, honey bees, plants, fish, aquatic invertebrates, and algae. Only 0.004—0.14 pounds of harpin protein per acre per season is reportedly necessary to protect crops and enhance yields. As with most proteins, harpin is a fragile molecule that is degraded rapidly by UV and natural microorganisms, and has no potential to bioaccumulate or contaminate surface or groundwater resources.

Deployment of harpin technology conserves resources and protects the environment by partially replacing many higher-risk products and enhancing the plant's own defense mechanisms, rather than introducing toxic biocides to the agricultural ecosystem. Decreasing the need to manufacture large volumes of pesticides also decreases the opportunity for terrorists to cause harm and panic by targeting pesticide production facilities.

4.15.1.1.2 Insect Growth Regulator in Bait System Greatly Lowers Hazards of Termite Control

Dow AgroSciences has developed a targeted approach to eliminating termite colonies that threaten structures, rather than the conventional approach of placing large volumes of insecticide into the soil surrounding a structure. Dow's Sentricon™ system, which has received U.S. EPA registration as a reduced-risk pesticide and by 2000 had been successfully applied on over 300,000 structures in the United States, significantly reduces the use of hazardous materials and the potential impacts on human health and the environment by attracting termites to a toxic bait station, rather than spreading insecticides throughout the soil.

The annual cost of termite treatments to the U.S. consumer is about $1.5 billion, and each year, as many as 1.5 million homeowners will experience a termite problem and seek a control option. From the 1940s until 1995, the nearly universal treatment approach for subterranean termite control involved the placement of large volumes of insecticide solutions into the soil surrounding a structure to create a chemical barrier through which termites could not penetrate.

A 2000-square feet home typically requires that 380 gallons of pesticide be pumped into the ground. A 100-home subdivision uses about 38 thousand gallons. Nationwide, pre-construction termiticide use is estimated at 400 million gallons.[67] Problems with this approach include difficulty in establishing an uninterrupted barrier in the vast array of soil and structural conditions, use of large volumes of insecticide, potential hazards associated with accidental misapplications, spills, off-target applications, and worker exposure. The inherent problems associated with the use of chemical barrier approaches for subterranean termite control created a need for a better approach, one of which is a termite bait toxicant.

The active ingredient in Sentricon™ is hexaflumuron, an insect growth regulator that interferes with termites' synthesis of chitin, the material that makes up the exoskeleton of insects. Although toxic to fish and not advisable for use where it could be washed out of the bait station into water, the potential for adverse effects on the ecosystem is dramatically reduced compared to previous methods because it is present only in very small quantities in stations with termite activity.

Hexaflumuron has very low toxicity to mammals[68] so its manufacturing facility would offer little interest to terrorists seeking explosive or hazardous targets. It is a replacement for the neurotoxic pesticide chlorpyrifos; a broad-spectrum chlorinated organophosphate that is the active ingredient in over 800 pesticide products and has been reported by the EPA Office of Pesticide Programs, Health Effects Division to be one of the leading causes of acute insecticide poisoning incidents in the U.S.

4.15.1.1.3 Microbial Insecticide Replaces Less Benign Synthetic Pesticides

Spinosad is a highly selective, environmentally friendly insecticide made by a soil microorganism, *Saccharopolyspora spinosa*, which was isolated from a Caribbean soil sample. Spinosad was developed by Dow AgroSciences and has proven effective in controlling many chewing insect pests in cotton, trees, fruits, vegetables, turf, and ornamentals, and displays relatively high selectivity in that 70—90% of beneficial insects and predatory wasps are left unharmed. Unlike traditional pesticides, it does not leach out, bioaccumulate, volatilize, or persist in the environment; it also has low toxicity to mammals and birds, thereby offering reduced risk to those who manufacture, handle, mix, and apply the product.

Controlling insect pests is essential to maintaining high agricultural productivity and minimizing monetary losses. Synthetic organic pesticides, from a relatively small number of chemical classes, play a leading role in pest control. The development of new and improved pesticides is necessitated by increased pest resistance to existing products, along with stricter environmental and toxicological regulations.

The microorganism makes spinosyns, unique macrocyclic lactones, containing a tetracyclic core to which two sugars are attached. Most of the insecticidal activity is due to a mixture of spinosyns A and D, commonly referred to as spinosad. Products such as Tracer® Naturalyte® Insect Control, and Precise® contain spinosad as the active ingredient. Insects exposed to spinosad exhibit classical symptoms of neurotoxicity: lack of coordination, prostration, tremors, and other involuntary muscle contractions leading to paralysis and death. Although the mode of action of spinosad is not fully understood, it appears to affect nicotinic and γ-aminobutyric acid receptor function through a novel mechanism.

Spinosad demonstrates low toxicity to mammals and other non-target organisms, and offers a replacement for many broad-spectrum insecticides, so its commercial development and availability helps to lower the amount of more hazardous pesticide produced, stored, transported, applied, and made vulnerable to terrorist-instigated accidents.

4.15.1.1.4 Microbial Fungicide Innocuous to Beneficial Insects

AgraQuest, Inc. has found a new biofungicide for fruits and vegetables based on a naturally occurring strain of bacteria. Sold under the name Serenade®, it is nontoxic to beneficial and non-target organisms, such as trout, quail, lady beetles, lacewings, parasitic wasps, earthworms, and honey bees. It does not generate any hazardous or synthetic chemical residues, and is safe to workers and groundwater.

Serenade® Biofungicide is based on a naturally occurring strain of Bacillus subtilis QST-713, discovered in a California orchard by AgraQuest scientists. The material works through a complex mode of action that is manifested both by the physiology of the bacteria and through the action of secondary metabolites produced by the bacteria. It prevents plant diseases, first, by covering the leaf surface and physically preventing attachment and penetration of the pathogens. In addition, it produces three groups of lipopeptides (iturins, agrastatins/plipastatins, and surfactins) that act in concert to destroy germ tubes and mycelium. The iturins and plipastatins have been reported to have antifungal properties. Strain QST-713 is the first strain reported to produce iturins, plipastatins, and surfactins, as well as two new compounds with a novel cyclic peptide moiety, the agrastatins. The surfactins have no activity on their own, but low levels (25 ppm or less) in combination with the iturins or the agrastatin/plipastatin group cause significant inhibition of spores and germ tubes. In addition, the agrastatins and iturins have synergistic activities towards inhibition of plant pathogen spores.

Serenade® has been registered for sale as a microbial pesticide in the United States since July 2000 and is also registered for use in Chile, Mexico, Costa Rica, and New Zealand, with registration pending in numerous other countries. The product has been tested on 30 crops in 20 countries and is registered for use in the United States on blueberries, cherries, cucurbits, grape vines, greenhouse vegetables, green beans, hops, leafy vegetables, mint, peanuts, peppers, pome fruit, potatoes, tomatoes, and walnuts, and is also registered for home and garden use.

It can be applied right up until harvest, providing needed pre- and post-harvest protection when there is weather conducive to disease development around harvest time. Available as a wettable powder, wettable granule, and aqueous suspension that is applied just like any other foliar fungicide, it can be applied alone or tank mixed. The wettable granule formulation is listed with the Organic Materials Review Institute (OMRI) for use in organic agriculture and will continue to be listed under the National Organic Standards, which were enacted in the United States in October 2002.

This fungicide's novel, complex mode of action, environmental friendliness towards non-target organisms, and relevance to a broad variety of fungal pests makes it an attractive replacement for more hazardous fungicides, including those that might constitute viable targets of terrorist activity.

4.15.1.1.5 Mimic of Insect Molting Hormone Is Toxic Only to Certain Insects

Rohm and Haas Company has developed CONFIRM™, a novel insecticide for controlling caterpillar pests in turf and a variety of agronomic crops that replace many less effective, more hazardous insecticides. It controls target insects through an entirely new mode of action that is inherently safer than current insecticides and represents a new class of insecticidal chemicals, the diacylhydrazines. These compounds strongly mimic a natural substance found within the insect's body called 20-hydroxy ecdysone, which is the natural insect molting hormone, and regulates development in insects. Because of this "ecdysonoid" mode of action, the new product powerfully disrupts the molting process in target insects, causing them to stop feeding shortly after exposure and to die soon thereafter.

Since 20-hydroxy ecdysone neither occurs nor has any biological function in most non-arthropods, CONFIRM™ is inherently safer than other insecticides to a wide range of non-target organisms such as mammals, birds, earthworms, plants, and various aquatic organisms. It is also remarkably safe to a wide range of key beneficial, predatory, and parasitic insects such as honeybees, lady beetles, parasitic wasps, predatory bugs, beetles, flies, and lacewings, as well as other predatory arthropods, such as spiders and predatory mites. Because of this unusual level of safety, the use of these products will not create an outbreak of target or secondary pests due to destruction of key natural predators or parasites in the local ecosystem. This should reduce the need for repeat applications of additional insecticides and reduce the overall chemical load on both the target crop and the local environment.

U.S. EPA has classified this product as a reduced-risk pesticide because it is inherently safer than other insecticides to a wide range of non-target organisms, poses no significant hazard to the applicator or the food chain, and does not present a significant spill hazard. In tests on mammals it has low toxicity by ingestion, inhalation, and topical application, and has been shown to be completely non-oncogenic, non-mutagenic, and without adverse reproductive effects.

Narrow-spectrum and organism-specific pesticides, such as those described in this section, help us control pests without resorting to conventional synthetic pesticides, which tend to be more general biocides, and have a greater harmful impact on humans and the environment. In turn, the presence of fewer hazardous pesticides in production, distribution, and widespread use results in a lower vulnerability to terrorists who might seek to exploit pesticides as a weapon against the human population.

4.16 ALTERNATIVE CLEANING TECHNOLOGIES

Substantial reductions in volumes of chemicals used can be achieved with new cleaning technologies based on replacement of chemicals used in traditional methods with substitutes proven to be effective. With many suitable alternatives becoming available, it is being said, for example, that the conventional use of vapor degreasing to clean electronics and other metal parts will soon be displaced and no longer used.[69] Whereas quick evaporation of volatile organic chemicals was desirable with the old vapor de-greasing methods, low volatility is key to the environmentally less harmful alternatives.

4.16.1.1.1 1,1,1-Trichloroethane
Ashland Chemical has introduced a line of substitutes for 1,1,1-trichloroethane that are blends based on hydrocarbon solvents and glycol ethers. While acknowledging that substituting one set of chemicals for another involves trade-offs that might include future restrictions on chemicals introduced as alternatives, an Ashland spokesman pointed to the overall desirable result of reducing chemical stocks. In the case where a company consumed 100 drums of 1,1,1-trichloroethane, less than half as much of the replacement (35–50 drums) would be sufficient to accomplish the same work.

Recycling and replenishment are also mechanisms reportedly used by Ashland Chemical to achieve quantitative minimization of chemical stocks and, thereby, an environmentally friendlier chemical process industry. Through collaboration with Shell Chemical, Ashland has introduced diacetone alcohol as a replacement for acetone in a cleaner used in ship building. Safer and longer lasting by virtue of its having a lower flash point and being one tenth as volatile, the diacetone alcohol cleaner is also offered with a complete service package. This means that the customer who purchases the new solvent can return it to Ashland Chemical after use for recycling and replenishment.

4.16.1.1.2 Perchloroethylene in Dry Cleaning
There are over 30,000 dry cleaning facilities in the United States, and approximately 95% of them use perchloroethylene (also known as PERC, or tetrachloroethylene) as the primary cleaning solvent. The National Institute of Environmental Health Sciences reports that short-term exposure to PERC can cause adverse health effects on the nervous system that include dizziness, fatigue, headaches, sweating, incoordination, and unconsciousness. Long-term exposure can cause liver and kidney damage. It is classified as a probable carcinogen. Most tetrachloroethylene is produced from ethylene via 1,2-dichloro-ethane, both of which are implicated in industrial accidents according to data from the EPA.[32]

Rynex Holdings Ltd. has developed and implemented an environmentally safe and effective dry cleaning method called RYNEX® that is composed of an azeotropic mixture of dipropylene glycol t-butyl ether (DPTB) and water. They claim this dry cleaning solution has low volatility and is non-flammable, non-carcinogenic, non-persistent in the environment, and economical to use. RYNEX® cleans both water-soluble and fatty acid stains using the same molecule, providing effective detergency and compatibility with existing dry cleaning technology for the effective removal of water and oil soluble stains without the shrinkage of wool fibers that occurs with wet cleaning methods. The product is now commercially available throughout the world.

Where adopted as a substitute for percholoethylene, this product will eliminate the risks associated with PERC, its manufacture, and its constituents, including the risk of explosions at facilities where these compounds are produced, used, or stored.

CONCLUSIONS

Efforts to develop the core tools of green chemistry predate the era when terrorism was a prominent concern, and were motivated primarily by environmental and economic concerns. Yet it has become abundantly apparent that many strategies to reduce environmental impact also turn out to offer the corollary benefit of reducing, or even eliminating, potent security risks. Hundreds of companies and facilities have innovated and implemented greener methods—such as those highlighted in this book—into their production and operation procedures, favorably impacting the health and safety of millions of Americans. Now proven, these same green methods are readily applicable to many thousands more facilities. Support for the on-going development and widespread adoption of green chemistry techniques therefore plays an important role in reaching the triple goals of environmental protection, economic competitiveness, accident prevention and national security against terrorist threats. We hope this book will assist decision-makers in industry and government who want to be a part of this eminently sensible effort.

REFERENCES

1. Baumann J, U.S. Public Interest Research Group Education Fund. *Protecting Our Hometowns, Preventing Chemical Terrorism in America: A Guide for Policymakers and Advocates* March 7, 2002 Available from: <http://uspirg.org/uspirg.asp?id2=5890>.

2. Hughart JL, Bashor MM. *Industrial chemicals and terrorism: human health threat analysis, mitigation and prevention.* Agency for Toxic Substances and Disease Registry; 1999.

3. Pianin E. Toxic Chemicals' Security Worries Officials in Washington Post 2001.

4. Orum P. *Chemical security 101: what you don't have can't leak, or be blown up by terrorists.* Center for American Progress; 2008.

5. Anastas PT, Kirchhoff MM. Origins, current status, and future challenges of green chemistry. *Acc Chem Res* 2002;**35**(9):686–94.

6. Anastas PT, Williamson TC, editors. Green chemistry: an overview. In: *Green chemistry: designing chemistry for the environment.* ACS Symposium Series 626. Washington, DC: American Chemical Society; 1996. p 1–17.

7. Anastas PT, Lankey RL. Life cycle assessment and green chemistry: the yin and yang of industrial ecology. *Green Chem* 2000;**2**(6):289–95.

8. Lin D, Mittelman A, Halpin V, Cannon D. *Inherently safer chemistry: a guide to current industrial processes to address high risk chemicals*, 20852. 3202 Tower Oaks Blvd, Rockville, MD: Technical Resources International, Inc.; 1994. p. 112.

9. Kletz TA. What you don't have, can't leak. *Chem Ind* 1978;287–92.

10. Anastas PT, Warner JC. First as paperback; originally published in 1998 ed. *Green chemistry: theory and practice*, 148. Oxford University Press; 2000.

11. Baird C. *Environmental chemistry.* 2nd ed. New York: W.H. Freeman; 1999.

12. ATSDR. *Benzene.* ATSDR Public Health Statement cited; 1999; Available from: <http://www.atsdr.cdc.gov/Toxprofiles/phs8803.html>.

13. Agency for Toxic Substances and Disease Registry, Industrial Chemicals and Terrorism: Human Health Threat Analysis, Mitigation and Prevention; 1999.

14. EPA. *Chemicals in the environment: Toluene.* cited 1999; Available from: <http://www.epa.gov/opptintr/chemfact/f_toluen.txt>.

15. Draths KM, Frost JW. Environmentally compatible synthesis of catechol from D-Glucose. *J Am Chem Soc* 1995;**117**(9):2395–400.

16. Niu W, Draths KM, Frost JW. Benzene-free synthesis of adipic acid. *Biotechnol Prog* 2002;**18**(2):201–11.

17. Li W, Xie D, Frost JW. Benzene-free synthesis of catechol: interfacing microbial and chemical catalysis. *J Am Chem Soc* 2005;**127**(9):2874–82.

18. Rogers RD, Seddon KR. *Ionic liquids as green solvents: progress and prospects*, vol. 856. Washington, DC: American Chemical Society; 2003.

19. Wang H, Gurau G, Rogers RD. Ionic liquid processing of cellulose. *Chem Soc Rev* 2012;**41**(4):1519–37.

20. Appel Y. Fuel depot bomb signals 'new phase of attacks', in *Irish examiner*, Associated Press: Ireland.

21. Reniers G, Herdewel D, Wybo J-L. A Threat Assessment Review Planning (TARP) decision flowchart for complex industrial areas. *J Loss Prev Process Ind* 2013;**26**(6):1662–9.

22. Committee on Assessing Vulnerabilities Related to the Nation's Chemical Infrastructure - National Research Council. *Terrorism and the chemical infrastructure: protecting people and reducing vulnerabilities*, vol. 152. Washington, DC: The National Academies Press; 2006.

23. Monterey Institute of International Studies Center for Nonproliferation Studies *Assessing Terrorist Motivations for Attacking Critical "Chemical" Infrastructure;* 2004: 253 pp.

24. Belke J. Chemical accident risks in U.S. industry – A preliminary analysis of accident risk data from U.S. hazardous facilities September 25, 2000, U.S. Environmental Protection Agency.

25. National Response Center. 2013; Available from: <http://www.nrc.uscg.mil/stats.html>.

26. Schierow L-J. Chemical Plant Security. Report for Congress received through the Congressional Research Service Web, January 23, 2003.

27. Gant SE, Atkinson GT. Dispersion of the vapour cloud in the Buncefield Incident. *Process Saf Environ Prot* 2011;**89**(6):391–403.

28. Johnson DM. The potential for vapour cloud explosions – Lessons from the Buncefield accident. *J Loss Prev Process Ind* 2010;**23**(6):921–7.

29. Orum P. *Preventing toxic terrorism: how some chemical facilities are removing danger to american communities*. Washington, DC: Center for American Progress; 2006. p. 44.

30. Leonnig CD, Hsu SS. Fearing Attack, Blue Plains Ceases Toxic Chemical Use in Washington Post November 10, 2001. p. A01.

31. Baumann J, Orum P. *Accidents waiting to happen: hazardous chemicals in the U.S. fifteen years after bhopal*. U.S. PIRG Education Fund and Working Group on Community Right-to-Know; 1999.

32. Kleindorfer PR, et al. *Accident history and offsite consequence data from RMP*Info*. The Wharton School, University of Pennsylvania, US Environmental Protection Agency; 2002. p. 46.

33. Center for the Study of Bioterrorism and Emerging Infections. *Chemical Terrorism Fact Sheet on Chlorine*; 2002; Available from: <https://erplan.net/WMD/ChemFiles/Links/ChemicalAgents/FactSheets/ChlorineFS.pdf>.

34. Reisch M, Johnson J. Safety derailed: After nine deaths, inspectors seek causes of deadly chlorine spill. p. 11.

35. General Accounting Office. *Wastewater Facilities: Experts' Views on How Federal Funds Should Be Spent to Improve Security*. Report to the Committee on Environment and Public Works, U.S. Senate January 2005; Available from: <http://www.gao.gov/new.items/d05165.pdf>.

36. Beyond Pesticides Daily News Archives Wastewater Plants Identified As Chemical Security Threat, Alternatives Advocated.

37. Kiely T, Donaldson D, Grube A. *Pesticide Industry Sales and Usage, 2000 and 2001 Market Estimates* 2004; Available from: <http://www.epa.gov/oppbead1/pestsales/01pestsales/market_estimates2001.pdf>.

38. New Jersey Department of Environmental Protection, Communication between Reggie Baldini and Paul Orum, Working Group on Community Right-to-Know, September 19, 2001.

39. *71 Fed. Reg. 388 (Jan. 4, 2006) and 71 Fed. Reg. 654 (Jan. 5, 2006).*

40. Toxics Use Reduction Institute. *UMass Lowell Institute Provides Technical Assistance to Eliminate Cyanide with Novel, Safer Industrial Process*. April 18, 2006; Available from: <http://www.turi.org/content/content/view/full/3713/>.

41. Suskind R. How an Al-Qaeda Cell Planned a Poison-Gas Attack on the N.Y. Subway, in Time 2006.

42. Sullivan A. *Their iGod*. The Daily Dish August 10, 2006; <http://time.blogs.com/daily_dish/2006/08/their_igod.html>.

43. Rotman D. Technology: Shell explores a cheaper route to MMA. *Chem Week* 1993.

44. Health and Safety Executive, Accident summary: release of hydrofluoric acid from marathon petroleum refinery, TX, USA, October 30, 1987.

45. Cusumano JA. New technology and the environment. *Chemtech* 1992;482–9.

46. Begley R. Hydrogen fluoride dodges a bullet. *Chem Week* 1993.

47. Sheldon RA. Consider the environmental quotient. *Chemtech* 1994;38–47.

48. Hunter D, Rotman D. Huls plans diisocyanate unit based on new phosgene-free process. *Chem Week* 1994.

49. Yagii T. *Manufacture of isocyanates without phosgene*. Japan: Daicel Chemical Industries; 1988.

50. Haggin J. Catalysis gains widening role in environmental protection. *Chem Eng News* 1994;22–30.

51. Sugano T. *Phosgene-free manufacture of high-molecular-weight polycarbonates*. Japan: Heisie; 1990.

52. Rogers C. *LanzaTech in Malaysian deal*, in *Taranaki Daily News*: New Plymouth, New Zealand.

53. International Occupational Safety and Health Information Centre. *Benzene*. May, 2003; Available from: <http://www.ilo.org/public/english/protection/safework/cis/products/icsc/dtasht/_icsc00/icsc0015.htm>.

54. Wikipedia, *Benzene*; 2006.

55. Duan W, Chen G, Ye Q, Chen Q. The situation of hazardous chemical accidents in China between 2000 and 2006. *J Hazard Mater* 2011;**186**(2–3):1489–94.

56. APELL (Awareness and Preparedness for Emergencies on a Local Level. Chinese River Contamination resulting from a petrochemical explosion and toxic spill: 5 dead, 70 wounded, 3.5 million without water. 24 November 2005; Available from: <http://www.uneptie.org/pc/apell/disasters/china_harbin/info.htm>.

57. Tanko JM, Blackert JF. Free-radical side-chain bromination of alkylaromatics in supercritical carbon dioxide (SC-CO$_2$). *Science* 1994;**263**(5144):203–5.

58. Mukhopadhyay T, Seebach D. Substitution of HMPT by the cyclic urea DMPU as a cosolvent for highly reactive nucleophiles and bases. *Helvetica Chimica Acta* 1982;**65**(39):38539.

59. Background Package from Pfizer Inc for September 13-14, 2004 *Psychopharmacological Drugs Advisory Committee of the Food and Drug Administration*, U.S. Food and Drug Administration.

60. Warner J, Cannon AS, Dye KM. Green Chemistry. *Environ Impact Assess Rev* 2004;**2004**(24):775–99.

61. Ryntz RA, Gunn VE. Pigments, paints, polymer coatings, lacquers, and printing inks. In: Kent J, editor. *Riegel's Handbook of Industrial Chemistry*. Netherlands: Springer; 1993; p. 1050–67.

62. *Powder Coatings Market - Global Industry Analysis, Size, Share, Growth, Trends and Forecast, 2012 - 2018*, Transparency Market Research (TMR).

63. Robin P, Lucisano LJ, Pearlswig DM. Rationale for selection of a single-pot manufacturing process using microwave/vacuum drying. *Pharm Technol* 1994;**18**(5):28–36.

64. Dagani R. Cuprate superconductors: record 133 K achieved with mercury. *Chem Eng News (USA)* 1993;**71**(19):4–5.

65. Mark Stevenson (Associated Press) Hundreds evacuated in central Mexico after pesticide leak, in The Boston Globe 9/13/2000.

66. Rohm and Haas Company Invention and commercialization of a new chemical family of insecticides exemplified by CONFIRM selective caterpillar control agent and the related selective insect control agents MACH 2 and INTREPID, a proposal submitted to the Presidential Green Chemistry Challenge Awards Program, 1998.

67. Beyond Pesticides. *Daily News Archive.* December 21, 2004; Available from: <http://www.beyondpesticides.org/news/daily_news_archive/2004/12_21b_04.htm>.

68. Beyond Pesticides Least Toxic Service Directory. *Safety Source for Pest Managment.* Available from: <http://www.beyondpesticides.org/infoservices/pesticidefactsheets/toxic/hexaflumuron.htm>.

69. Kirschner EM. Environment, health concerns force shift in use of organic solvents. *Chem Eng News (United States)* 1994;**72**(25).

Printed in the United States
By Bookmasters